THE SPACE INDUSTRY

TRADE RELATED ISSUES

ORGANISATION FOR ECONOMIC CO-OPERATION AND DEVELOPMENT

Pursuant to article 1 of the Convention signed in Paris on 14th December, 1960, and which came into force on 30th September, 1961, the Organisation for Economic Co-operation and Development (OECD) shall promote policies designed:

- to achieve the highest sustainable economic growth and employment and a rising standard of living in Member countries, while maintaining financial stability, and thus to contribute to the development of the world economy;
- to contribute to sound economic expansion in Member as well as non-member countries in the process of economic development; and
- to contribute to the expansion of world trade on a multilateral, non-discriminatory basis in accordance with international obligations.

The Signatories of the Convention on the OECD are Austria, Belgium, Canada, Denmark, France, the Federal Republic of Germany, Greece, Iceland, Ireland, Italy, Luxembourg, the Netherlands, Norway, Portugal, Spain, Sweden, Switzerland, Turkey, the United Kingdom and the United States. The following countries acceded subsequently to this Convention (the dates are those on which the instruments of accession were deposited): Japan (28th April, 1964), Finland (28th January, 1969), Australia (7th June, 1971) and New Zealand (29th May, 1973).

The Socialist Federal Republic of Yugoslavia takes part in certain work of the OECD (agreement of 28th October, 1961).

Publié en français sous le titre:

L'INDUSTRIE SPATIALE
Questions liées aux échanges

This study was undertaken within the general framework of the OECD's joint work on issues in trade in high technology products by the Industry Committee and the Committee for Scientific and Technological Policy. It is one of four case studies in high technology industries undertaken in this context, the others concerning the machine-tool, pharmaceutical and semi-conductor industries. The study was prepared by a member of the OECD Secretariat, Mr. Patrick Dubarle assisted by a consultant, Ms. Susan Becker, and under the general direction of an ad hoc Group of Government experts. The study was examined by the Industry Committee and the Committee for Scientific and Technological Policy and was derestricted under the responsability of the Secretary-General of the OECD.

Also available

THE SEMICONDUCTOR INDUSTRY. Trade Related Issues (February 1985)
(93 85 01 1) ISBN 92-64-12687-2   146 pages   £9.50   US$19.00   F95.00

THE PHARMACEUTICAL INDUSTRY. Trade Related Issues (September 1985)
(93 85 03 1) ISBN 92-64-12737-2   £4.50   US$9.00   F45.00

Prices charged at the OECD Publications Office.

THE OECD CATALOGUE OF PUBLICATIONS and supplements will be sent free of charge
on request addressed either to OECD Publications Office,
2, rue André-Pascal, 75775 PARIS CEDEX 16, or to the OECD Sales Agent in your country.

TABLE OF CONTENTS

Contents

Chapter I

INTRODUCTION

As the governments of OECD Member countries become increasingly aware of the benefits firms can derive from mastering new technologies amidst international competition, they are endeavouring to make these industries more competitive by enhancing the effectiveness of industrial and technological policies. This has been reflected in the higher priority given to high technology and the wider range of measures designed to further expansion of these sectors.

As these trends increase, they create fears that government intervention can be excessive or too specific. Particular attention is being paid to the international implications of such intervention. The principal question is whether increasing government involvement with high technology products will not create distortions in competition and lead to friction at an international level. These were the considerations underlying the request made to the OECD for a study of this subject.

In this context the choice of a case study on the space industry is especially interesting. First, due to the particularly strong emphasis on R&D, the experimental character of many of these products, and the highly qualified personnel employed to build, launch and orbit satellites, this industry is the spearhead of technological developments. Second, spin-offs from space programmes extend to the manufacturing industry as a whole and the services sector through inter-sectoral transfers of technology. Space activities provide an important outlet for many industries: telecommunications, computers, electronics, chemical combustion motors, scientific instruments, etc., and are therefore a source of much innovation. Third, the role of government has been essential in laying the industry's foundations and continues to be important. Fourth, private sector trade has made little contribution as yet to the industry's activities.

However, as commercialisation becomes more prevalent in many space areas, the volume of trade in space items and services is increasing, favoured by the broadening of the supplier base and growing international co-operation. This co-operation is particularly reflected in the growth of subsystem and component trade. A multipolar trading system is gradually developing, although it is still limited in scope.

In view of the present inadequacy of national statistical collection offices to supply appropriate trade data, the analysis will focus mainly on how markets operate and on subsequent changes. As regards the space industry, the trade situation results to a small extent from the existence of direct barriers to trade *per se*. It is mainly a consequence of the present structure of competition and markets. The former are influenced by factors such as policies of national preference, tendering procedures of international bodies, the development of programmes in co-operation and the often decisive contribution of military procurement and export financing.

Chapter II

SPACE ACTIVITIES

1. BACKGROUND

Although the conquest of space is a fairly old idea, no notable scientific advances in astronautics occurred before the end of the 19th century (Tsiolkovski) and the beginning of the 20th (Goddard, Oberth). At that time the principle of the multi-stage rocket was established as a means of escaping the Earth's gravity. The first important experiments were achieved after World War I[1] and significant civil technical successes appeared slowly despite the spur to research following World War II. The first artificial earth satellite, SPUTNIK, was not launched by the Soviet Union until the late 1950s.

The 1960s

Stimulated by competition with the Soviet Union, which achieved the first manned orbital flight in 1961, the United States commitment to space exploration became clear under John F. Kennedy's leadership. His April 1961 speech established a national target of a moon landing by the end of the decade.

The underlying space policy principles were then the development of the United States' technological lead, intensification of its general scientific efforts to open new frontiers of knowledge and strengthening of national security. Economic and social criteria were also taken into account as well as the need to develop space projects for the benefit of all mankind.

Under government prompting, appropriations for the National Aeronautics and Space Administration, NASA, formed in 1958, increased regularly until 1965 (see Graph 1). Large-scale programmes (GEMINI, APOLLO) were started and basic space flight techniques (precise orbital placement, rendezvous, docking, etc.) were developed and tested.

Simultaneously with the moon programme, scientific missions (earth observation and planetary exploration using unmanned spacecraft) and military missions developed. The main initial spin-offs from space projects were in satellite developments (communications, meteorology, navigation, geodesy, etc.). Thanks to the work of NASA, satellite technology developed rapidly from passive satellites, only reflecting waves, such as ECHO, to active satellites, relaying and amplifying signals, such as RELAY. The culmination was achieved in 1964 with the launch of the first geostationary telecommunications satellite, SYNCOM 3[2],

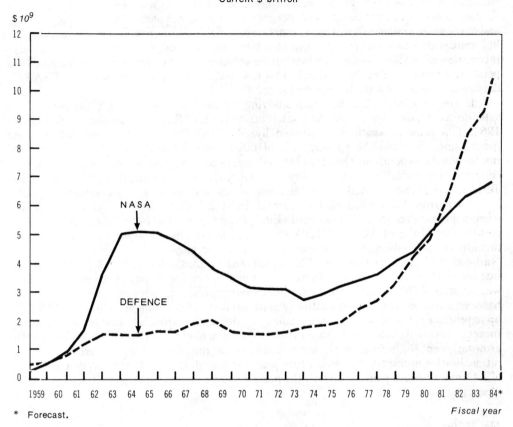

Graph 1 **UNITED STATES SPACE EXPENDITURES, 1959-1982**
Current $ billion

$ 10^9

NASA

DEFENCE

1959 60 61 62 63 64 65 66 67 68 69 70 71 72 73 74 75 76 77 78 79 80 81 82 83 84*

* Forecast.

Fiscal year

Sources : Futures 82 (based on data from the Office of Management and budget) ; OECD Secretariat.

when the United States, through a public corporation (Comsat), and ten other nations contributed to the foundation of an international organisation, INTELSAT (August 1964), to develop and manage an international satellite telecommunications network.

At the same time, the United States space programme placed considerable emphasis on openess and sharing of benefits which led to an extensive network of international co-operation. In many cases this co-operation provided a stimulus to other countries' space programmes. For example NASA and the United Kingdom co-operated in the launch of ARIEL 1 in 1962. During the same year, Brazil, France, Italy and the United Kingdom built earth stations to communicate with NASA's SYNCOM.

Throughout this period the role of the military sector was noted primarily in propulsion technology. The JUPITER C, a US Army rocket, launched the first US satellite, EXPLORER. Starting with the Eisenhower administration, however, the US has insisted upon maintaining separate civilian and military sectors. Although this has never ruled out co-ordination on specific projects, it has been maintained ever since.

Although the Defense Department commissioned or carried out a substantial volume of research, NASA was the largest investor in space R&D. This was NASA's main function because it was required to hand over experimental programmes once they became operational and marketable and therefore it did not maintain its own construction or fabrication facilities, but continued to own and operate substantial laboratory and testing facilities. Construction or fabrication of NASA funded satellites, space vehicles, rockets and control equipments was performed under contract by industry. This has resulted in the transfer of some of NASA's developed technologies to the private sector.

During this time, Canada began studying the earth's ionosphere and the potential of satellite communications. Their first satellite, ALOUETTE I, was launched in September 1962. This project, which was followed by ALOUETTE II and the two ISIS satellite programmes, involved the participation of other countries and set a lasting pattern of international co-operation. The data obtained were used to improve worldwide radio links and thus proved that satellites were a feasible solution to many communication problems. In 1969 Telesat Canada was created to provide domestic satellite services on a commercial basis.

Apart from the United States, Canada and the USSR, the other governments that showed interest in space activities were in Europe and Japan. The United Kingdom funded the construction of the BLACK ARROW rocket which put a device into orbit in 1971. Public institutions and government agencies were set up, notably in France, where the Centre National d'Etudes Spatiales (CNES) which was established in 1962, launched its first satellite in 1965: and, in Japan where the National Space Development Agency (NASDA) was set up in 1969[3]. Two European multinational organisations were set up in 1964. ELDO, based on a United Kingdom suggestion, was meant to develop a launcher while ESRO was set up to promote space research and technology by developing artificial satellites and sounding rocket payloads. Technical difficulties in the development of the EUROPA launcher and the growing need to include application satellites in the development programme made rationalisation necessary. In 1966, a European space conference led to the gradual definition of a European space policy, and was the first step towards establishing the European Space Agency (ESA).

The 1970s

The successful moon landing of a manned spacecraft in 1969 (APOLLO 11) confirmed the technological lead of the United States, but was followed by a period of relative budgetary restraints for NASA. Less sensitive than before to arguments about prestige, and more concerned about the economic costs rather than the scientific yields of space investments, the American Congress, and to some extent the Executive branch, imposed a slow-down on the final programmes of the 1960s (particularly APOLLO and SKYLAB). There were no American manned flights between 1975 and 1981 and new projects were severely pruned, especially as the rapidly growing inflation limited NASA's financial capabilities.

While the USSR carried out a programme of frequent launches and the testing of orbital space stations (SALYUT), cargo ships or docking ports (PROGRESS), the United States gave priority to the construction of a reusable space vehicle, the Space Shuttle. Meanwhile, certain experimental satellite programmes were continued such as the ATS – Applications Technology Satellites – for telecommunications. Others, such as the LANDSAT series, were undertaken for remote sensing. The development of geostationary weather satellites supported numerous World Meteorological Organisation (WMO) programmes to improve weather forecasting and storm warnings in the western hemisphere. Several scientific research spacecraft, including VOYAGER 1 and 2, were also launched to the outer planets.

During the 1970s this trend in the United States reflected a change in policy. Previously, market forces only played a minor role and the building of space capabilities was motivated primarily by decisions made on behalf of the government agencies concerned. Moreover, space expenditures were considered a way of financing innovation, due to the possibilities offered to industries by the dissemination and use of space research[4] to develop new products and improve manufacturing processes[5]. With the reduction of some of the United States' prestigious scientific programmes, attitudes changed leading to a greater interest in developing commercial applications of experimental space projects.

This change was naturally accompanied by a reduction in NASA's budget (at least up to 1975). In the space telecommunications area, this trend had started even earlier. NASA paved the way for the commercialisation of basic satellite communication technology. First, the pattern of NASA sponsored international co-operation with the users of international systems considerably accelerated the establishment of INTELSAT. Second, NASA contracts induced the creation of hardware and system expertise in a large variety of firms such as Hughes, RCA, etc. Taking advantage of this transfer of technology, Hughes was able to supply the first commercial user, INTELSAT, with Early Bird in 1965. From then on, the pace of development of the US satellite supplier industry was fixed by user demand, such as military institutions and international organisations. Demand from private common carriers followed albeit with a certain delay during the 70s. Table 3 and Annex I.1 clearly illustrate this shift from the public to the private sector for non-military satellites.

In Canada, ANIK A1, launched in November 1972, was the world's first domestic geosynchronous communications satellite system carrying telephone and television traffic. The Canadian-built HERMES, (the Communications Technology Satellite, launched in January 1976 as a co-operative project with NASA and ESA), pioneered new technologies (i.e. Ku-band and 200 watt transponders) and satellite applications (i.e. direct broadcasting, tele-education, tele-health). ANIK B, the first hybrid satellite, launched in December 1978, permitted commercial communications in the C-band and experimental work in the Ku-band.

Outside the United States and Canada, the market economy hesitated or failed to get involved in space activities. Because of public monopolies in telecommunications, user demand was channeled through the PTT, where interest in space communications emerged slowly. Europe was estimated to be ten years behind the United States in the satellite field. As manufacturers went through phases of experimental product development and gradual refining of basic satellite and launcher technologies, the foreign licensing of US technologies to these countries helped the building of indigenous capacities. This began in 1967 and the first achievement came in 1974 and 1975 with the launch of the French/German experimental telecommunications satellites SYMPHONIE. By the end of the decade the technological gap was largely made up.

In 1975 the European Space Agency (ESA) was established by a merger of ELDO and ESRO[6]. The current ESA Member states are Belgium, Denmark, France, Federal Republic of Germany, Ireland, Italy, Netherlands, Spain, Sweden, Switzerland and the United Kingdom. Austria and Norway have associate Member status. Since 1979 Canada has co-operated closely with ESA under an Agreement of Co-operation. ESA expenditures have grown annually since its foundation despite a slight decrease in constant price until 1982 (see Table 1 and Annex I.2). ESA has been instrumental in implementing several major European programmes such as the ARIANE launcher and SPACELAB, and in supporting a major telecommunications satellite programme. These developments have not however jeopardised the maintenance and pursuit of several extensive national projects. In Japan, trends in the NASDA budget have been similar to ESA spending (see Table 2).

11

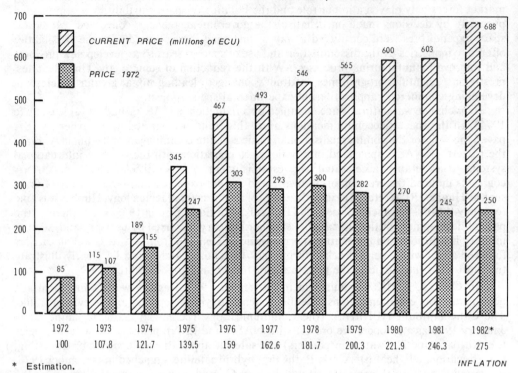

Table 1 **TREND IN ESA BUDGET, 1972-1982**

Legend:
- CURRENT PRICE (millions of ECU)
- PRICE 1972

	1972	1973	1974	1975	1976	1977	1978	1979	1980	1981	1982*
Current price	85	115	189	345	467	493	546	565	600	603	688
Price 1972	85	107	155	247	303	293	300	282	270	245	250
INFLATION	100	107.8	121.7	139.5	159	162.6	181.7	200.3	221.9	246.3	275

* Estimation.

Source : ESA Budget.

By the end of the 1970s two trends were clear:

− First, continued internationalisation of the space sector, as mentioned above, with the strengthening of the Canadian and the European commitment and reinforced through the consolidation of multinational organisation activities [the range of INTELSAT satellites grew in weight and power[7], see Annex I.3]. Along these lines, new regional satellite communication organisations, e.g. EUTELSAT, ARABSAT and specialised bodies such as INMARSAT, EUMETSAT, were formed to meet user requirements in particular markets. The enlarging of the club of countries (including Indonesia, China, India and Brazil) that have or intend to put satellites into orbit is another indication of internationalisation trends (see Annexes I.4 and I.5). With the increasing awareness of the benefits to be derived from space, many countries using telecommunications services and remote sensing data are putting greater pressure on international bodies such as the United Nations and the International Telecommunication Union to obtain advantages in these markets. Developing countries are more insistent in their claims for reserved slots in the geostationary orbit. Another concern is for prior consent to the right to sell observation data about their resources, and to control foreign television broadcasting into their territory.

12

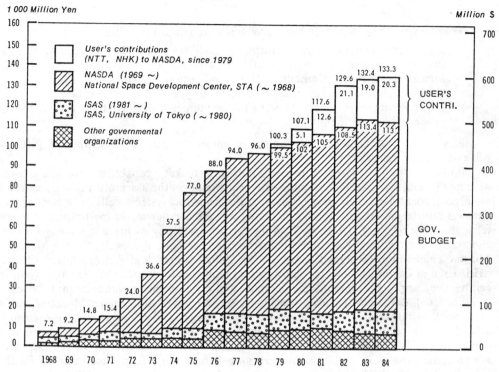

Table 2   TREND AND BREAKDOWN OF SPACE BUDGET IN JAPAN

1 000 Million Yen   *Million $*

Legend:
- □ User's contributions (NTT, NHK) to NASDA, since 1979
- ▨ NASDA (1969 ~) / National Space Development Center, STA (~ 1968)
- ▦ ISAS (1981 ~) / ISAS, University of Tokyo (~ 1980)
- ▩ Other governmental organizations

USER'S CONTRI.

GOV. BUDGET

1968 69 70 71 72 73 74 75 76 77 78 79 80 81 82 83 84

Note : US $ approx. 240 yen.   *Japan fiscal year*
Source : NASDA.

– Second, growing defence outlays for the development of new technologies and innovation in the majority of the countries competing in the aerospace sector. In the United States, the military space budget was less than half NASA's budget in 1970, but had almost equaled it by 1980 and exceeded it by 1982 (see Graph 1). Defence R&D expenditures have accelerated, especially in countries producing missiles. Although orbital and ballistic techniques now show more pronounced divergences, the distinction between civilian and military technology remains less clear-cut in certain domains than in the past. In the United States, the national Space Transportation System, i.e. the Space Shuttle, flies for the Defense Department and NASA, as well as for foreign and domestic civilian users. Civilian satellites are sometimes also used by military authorities.

To give an idea of the pattern of space applications between 1975 and 1980, 695 (80 per cent of the 922 launched payloads worldwide) were Soviet military satellites[8]; 38 per cent of the non-military payloads were for telecommunications; 20 per cent for scientific missions; 13 per cent for observation; 12 per cent for weather forecasting and 4 per cent for navigation.

13

2. SPACE PRODUCTS, HARDWARE AND SERVICES

Traditionally there are three basic production activities within this sector:

– the development and manufacturing of satellites which differ according to their function: telecommunications, earth observation and remote sensing, weather forecasting, and relays among satellites and between Earth[9];
– the manufacture and launching of vehicles to put satellites into orbit;
– the construction of ground stations to communicate with satellites and/or receive information from them.

Production processes for this equipment involve a complex assembly of systems, sub-systems and components.

Building a launcher (i.e. at present a multi-stage rocket[10] capable of putting a satellite weighing several hundred kilogrammes into orbit) requires the assembly and integration of metal components (structure, nose cone), propulsion and fuel systems, cells and miscellaneous hardware (turbo pumps, instruments including onboard computers, the inertial guidance and telemetry systems, etc.). These are manufactured respectively by firms in the engineering, chemicals, electrical and electronics industries; for example, Rockwell, Martin Marietta, Morton Thiokol and General Electric in the United States, and Aerospatiale, SEP and MBB-Erno in Europe. Launching logistics also require a terrestrial infrastructure (launch centre, electronic flight monitoring equipment and comprehensive management services).

For satellites (i.e. the space segment) there are a variety of sub-systems. Most consist of a platform[11] (metal skeletons and panels, solar panels, apogee and in some cases perigee motors, attitude control, batteries, etc.) and a payload (onboard electronics for frequency amplification, modulation/demodulation and multiplexing, receiver and transmitter aerials, etc.). For the latter, the main contractor is often an electronics firm (e.g. Thomson-CSF for the French satellite TELECOM 1) whereas for the platform or the entire satellite the project leader is usually an aerospace firm (e.g. Ford Aerospace for the INTELSAT V series, Aerospatiale for ARABSAT, Hughes Aircraft for INTELSAT VI, Spar Aerospace for ANIK D and BRASILSAT).

Lastly, the earth station network (i.e. the ground segment) involves similar telecommunications equipment: receiver/transmission antenna (though with very different dimensions, especially for main stations) together with aerial mobility devices, optical receiver systems, computer/telecommunications and radio electrical equipment, etc. These items are manufactured by telecommunications or electronics companies (e.g. NEC or Fujitsu in Japan, ITT in the United States, Marconi and Thomson-CSF in Europe, AEL Microtel, Raytheon and Spar in Canada).

No description of the space sector would be complete without mentioning the various ancillary services required for successful technological applications. These services are burgeoning vigorously in North America and now in Europe: they mainly pertain to:

– launch services. To prepare for the flight of the rocket, monitor it and guide the space craft into its pre-defined trajectory. This was previously the responsibility of the space agencies, NASA, ESA, CNES, but is gradually being commercialised. However, these agencies continue to carry partial or entire infrastructural costs (site, ancillary equipment). In Europe, ESA created, as early as 1980, Arianespace, an organisation functioning on a commercial basis to take over the production, management and commercialisation of the ARIANE rocket. In the United States, under the NASA ELV commercialisation procedure, operating and manufacturing rights are being

14

transferred to a launch service company, but the launch vehicles remain the property of the Agency. NASA continues to operate the Space Shuttle, which remains the primary US launch vehicle.

- telecommunications services. With advancing deregulation in the United States, private sector intermediaries come between telecommunications users and satellite producers such as the telecommunications network managers, i.e. who lease or own the transponders (systems for amplifying and retransmitting information received)[12]. As regards the satellite, ownership is diversified and depends on the carrier (so-called by the analogy with the communications field prior to the space age). Ownership may be shared by an international organisation, such as INTELSAT, a regional organisation, such as EUTELSAT[13], by various countries' PTTs or by public companies. Satellites may also be owned by private service companies for national networks;

- consulting, management and training services which are becoming increasingly important (for example, Comsat General of the United States and the Satel-Conseil group in France).

Chapter III

SPACE MARKETS AND DEMAND

At present (the mid 80s), space is considered an emerging industry whose outlets still depend considerably on public sector markets. For some years, however, great attention has been paid to the spin-offs of expenditures committed to space programmes and on how to involve the private sector. New markets have been opened to competition, especially in services, and mainly in the United States under the influence of deregulation policies. However policies of national or regional preference remain deep-rooted and progress towards the continued opening of markets for international trade has been hesitant.

As space developments accelerate (150 to 200 civil satellites will be launched onwards to 1990 in the world outside the Eastern Block)[14] the markets are undergoing substantial structural change. The communications industry with its corresponding services is the guiding force and private demand is growing, though with important differences according to the country. To illustrate these developments and put them in perspective, we shall look successively at:

1. the market for satellite telecommunications and related services which is growing steadily and offers several challenges;
2. the market for observation and weather forecasting satellites: a captive market for which the countries involved are contemplating commercialising their products;
3. the market for earth stations, which has fairly severe competition and a diversified customer base.

In view of the very special features of the market for launchers and launch services, we have opted to take the above aspects into consideration in the following chapter on prices, costs and competition, thereby emphasizing the key economic role of launching, i.e. satellite transport.

1. TELECOMMUNICATIONS SATELLITES AND RELATED SERVICES

Due to the technical advantages of satellites, a genuine market has opened up. First, compared with coaxial cables or wireless links, satellite transmission offers a flexible, competitive answer to long-distance communications problems between two points, from one point to several points and linkages between moving points (ships, aircrafts). A satellite can also provide top quality transmission for an area with a low population density and few telephone lines (e.g. many developing countries) and make the handling of large information flows possible within short periods of time. Next, the related technologies have proved to be reliable, while continuous progress has been achieved regarding costs. Lastly, it has been necessary to shift towards satellites to satisfy growing needs for telephony[15] (an annual increase of 6 per cent in telephone installations worldwide during the 1970s and probably a far higher increase in the industrialised countries) to avoid the risk, at least temporarily, of saturating conventional networks.

In these circumstances, there was great potential for growth in the demand for telecommunications satellites; but even so, the market took off fairly slowly. Between 1970 and 1975 there were 17 civil launches worldwide, (Western and third world only), and 20 during the following five years (see Table 3). The in-orbit transponder capacity, however, grew substantially as the ratio of transmission capacity per satellite increased.

In the United States, the 1973 decision to end the experimental ATS programme meant that NASA moved aside and that the private sector henceforth formed the bulk of civilian market orders; this also applied to Canada. In Europe, conversely, demand continued to be channelled through ESA or national PTTs. Direct private sector investment has developed relatively recently, for example MERCURY and UNISAT in the United Kingdom, CTL and CORONET in Luxembourg.

The market is now taking off more than fast enough to double capacity every four years (see Table 4 and Annex I.6). Moreover, an average of 23 geostationary satellites are expected to be launched each year until 1990, i.e. approximately eight US DOMSAT, four international satellites (INTELSAT and INMARSAT) and 11 domestic and regional satellites (centrally planned economies excluded)[16]. Related 1983-1990 investments will amount to about $4 billion for launches, $4.5 billion for satellites and $8 billion for earth stations.

This growth in demand[17] is accompanied by:

i) stronger competition for launch services, resulting in lower user costs. With US expendable launch vehicles, launching prices for an INTELSAT Class V Satellite (see Annex I.3) have been in the neighbourhood of $50 – $60 million. Newer launching systems brought this down appreciably[18]. Space transportation costs could also be lowered by the use of reusable launchers such as the Space Shuttle, but opinions are still divided on this (see Chapter IV).

ii) larger variety and versatility of satellites examplified by their greater weight (e.g. well in excess of one ton for the OLYMPUS and INTELSAT VI platforms) and volume. These improvements were due to increased launching capability and advances in satellite technology. Satellites can include increasingly complex equipment which improve their output, margin for manoeuvrability (thanks to additional fuel storage capacity) and above all, their adaptability to a greater variety of possible uses. By increasing the power and flow rate of information transmitted, not only can telephone links and radio/television broadcasting be relayed but, more broadly, any type of digital communication can be transmitted from one point to another. Examples are data transmission between computers, newspapers, facsimile transmission, video conferencing, videotex, etc.[19].

iii) development of new services. These services include tele-education and tele-care; each uses satellite links and small earth stations to provide services to isolated areas. Tele-education lets instructors in more developed areas transmit information to people in less accessible regions. Tele-care enables doctors and health care personnel in smaller, often more remote, care centers to communicate directly with medical professionals at major hospitals and scientific institutions.

Nevertheless, the growth of the telecommunications services market is irregular as each newly launched satellite increases minimum capacity. The supply side advances with successive generations at the pace of replacement, according to system lifespans. Thus, over-capacity develops at the beginning of a cycle, to be gradually absorbed as demand increases. Accordingly, the present situation is expected to put strain upon the supply side by the end of the decade[20].

Year launched	United States		Europe	
	Public R & D	Commercial	Public R & D	Commercial
1958				
1960	ECHO			
	ECHO I			
1961		OSCAR		
1962	RELAY I	TELSTAR		
	ECHO AVT-1			
	ECHO AVT-2			
1963	SYNCOM I	TELSTAR		
	SYNCOM II			
1964	ECHO II			
	RELAY II			
	SYNCOM III			
1965				
1966	ATS-I			
1967	ATS-II			
	ATS-III			
1968	ATS-IV			
1969	ATS-V			
1970				
1971				
1972				
1973				
1974	ATS-VI	WESTAR A	SYMPHONIE A	
		WESTAR B	(France/Germany)	
1975		RCA-A	SYMPHONIE B	
			(France/Germany)	
1976		RCA-B		
		COMSTAR A		
		COMSTAR B		
		MARISAT A		
		MARISAT B		
		MARISAT C		

d People's Republic of China

Other countries		International	US/NATO
Public R & D	Commercial	Commercial	Military
			SCORE
			COURIER
OUETTE I			WESTFORD
DN)			EXPERIMENT
			WESTFORD
			EXPERIMENT
OUETTE II		INTELSAT I	LES-1
DN)		(EARLY BIRD)	LES-2
			LES-3 & 4
		INTELSAT II(F-1)	DSCS PHASE I
			(26 SAT.)
			LES-5
		INTELSAT II (F-2)	
		INTELSAT II (F-3)	
		INTELSAT II (F-4)	
		INTELSAT III (F1)	LES-6
		INTELSAT III (F-2)	
IS I		INTELSAT III (F-3)	TACSAT
DN)			SKYNET 1
		INTELSAT III (F-4)	
		INTELSAT III (F-5)	
		INTELSAT III (F-6)	NATO-A
		INTELSAT III (F-7)	
		INTELSAT III (F-8)	
S II		INTELSAT IV (F-2)	NATO-B
DN)		INTELSAT IV (F-3)	DSCS II (2 SAT.)
	TELESAT-1/ANIK-A1	INTELSAT IV (F-4)	
	(Canada)	INTELSAT IV (F-5)	
	TELESAT-2/ANIK-A2	INTELSAT IV (F-7)	DSCS II (2 SAT.)
	(Canada)		
		INTELSAT IV (F-8)	SKYNET 2A
			SKYNET 2B
	TELESAT-3/ANIK-A3	INTELSAT IV (F-6)	DSCS II (2 SAT.)
	(Canada)	INTELSAT IV (F-1)	
RMES		INTELSAT IVA (F-1)	
nada)	PALAPA-A	INTELSAT IVA (F-2)	NATO III-A
nt project	(Indonesia)		LES 8.9
A/CDN/US)			

Year launched	United States		Europe	
	Public R & D	Commercial	Public R & D	Commercial
1977			SIRIO (Italy) OTS A (ESA)	
1978		COMSTAR-D3	OTS B (ESA)	
1979		WESTAR 3		
1980		SBS 1		
1981		COMSTAR 4 SBS 2 RCA SATCOM 3		MARECS 1
1982		RCA SATCOM 4 WESTAR 4 WESTAR 5 RCA SATCOM 5 SBS 3		
1983	TDRS A	SATCOM 1R GALAXY 1 TELSTAR 301 SATCOM 2R GALAXY 2		ECS 1(ESA)
1984		SPACENET F1 GALAXY 3 SPACENET F2 SBS 4 TELSTAR 302		ECS 2 (ESA) TELECOM 1A (France) MARECS B2
1985 (feb.)				

Sources: *Futuribles,* November 1980 and OECD Secretariat.

d People's Republic of China *(cont'd)*

Other countries		International	US/NATO
Public R & D	Commercial	Commercial	Military
TS-II (Japan)	PALAPA B	INTELSAT IVA (F-4)	NATO III-B
S (Japan)	(Indonesia)	INTELSAT IVA (F-5)	DSCS II (2 SAT.)
S (Japan)	TELESAT-4-ANIK-B	INTELSAT IVA (F-3)	NATO III-C
	(Canada)	INTELSAT IVA (F-6)	FLEETSATCOM
			DSCS II (2 SAT.)
			NAVSTAR (3 SAT.)
			DSCS II (2 SAT.)
			FLEETSATCOM
		INTELSAT V (F-2)	FLEETSATCOM(2SAT.)
			NAVSTAR (2 SAT.)
PPLE (India)		INTELSAT V (F-1)	TRANSIT NOVA
TS IV (Japan)		INTELSAT V (F-3)	NAVSTAR
			FLEETSATCOM
	INSAT 1A (India)	INTELSAT V (F-4)	DSCS III (2 SAT.)
		INTELSAT V (F-5)	
TS III (Japan)	TELESAT-5/ANIK D1		
	TELESAT-6/ANIK C1		
	TELESAT-7/ANIK C2		NAVSTAR
	(Canada)		
S 2A (Japan)	PALAPA B1	INTELSAT V (F-7)	DSCS III
S 2B (Japan)	(Indonesia)		
	TELESAT-8/ANIK C3		
	(Canada)		
	BS/2A (Japan)	INTELSAT V (F-8)	SYNCOM IV 1 (LEASAT)
	TELESAT-9/ANIK D2		SYNCOM IV 2
			NATO III D
			TRANSIT NOVA
	ARABSAT 1		SIGINT
	(Arab League)		
	BRASILSAT (Brazil)		

Table 4

World market trend for telecommunications system space segment

Transponders in orbit – equivalent 36 MHz

	1972	1976	1982	1984	1990[1]
INTELSAT	100	180	360	510	750
North American	12	146	310	520	1 125
Rest of world	0	24	70	100	625
(including Europe				(30)	(200)
Total	112	350	740	1 130	2 500

1. Forecast estimate, excluding military systems, satellites nearing the end of their lifespan, including reserve capacities, and excluding USSR and China.
Source: SEST/Euroconsult, 1984-85.

Global systems

Presently, INTELSAT[21] which provides about ⅔ of intercontinental telephone traffic, accounts for the major share of telecommunications demand originating with multinational organisations. The organisation's circuits demand will probably continue to grow at a sustained annual rate of about 15 per cent for several years. Construction contracts have been awarded for the first group of five satellites in the VI series (to be launched in 1986 and 1987) corresponding to a turnover of $0.52 billion. It is estimated, that for the period 1988-1991, increases in international traffic and replacement demands could be met in terms of capacity and location by four class V-A satellites and six more satellites of the same size and capacity as INTELSAT VI.

Excess capacity is expected until 1986 and maybe beyond depending on the pattern of utilisation of the VI series. Thus INTELSAT now leases transponders solely for domestic requirements. At the end of 1983, 32 countries using 48 transponders were benefiting from this system (see Annex I.7). The Organisation is also interested in the development of new video networks (through full-time dedicated international TV leases) and in the market for mobile telecommunications[22]. Since October 1983, INTELSAT has provided a new range of services tailored to business uses called Intelsat Business Services (IBS)[23].

These activities reflect efforts to cope with new competitors in world markets. Telecommunications operators are now increasingly inclined to lease domestic excess capacities to foreign customers, as illustrated by the extension of United States domestic systems into neighbouring countries and the international development of business service networks on a private basis[24]. Furthermore other systems (ARABSAT, EUTELSAT) are offering services particularly suited to regional needs because *(a)* satellites can be placed in orbit so that their power is optimally concentrated and utilised; *(b)* they can be in a location which adequately accesses the geographic area to be serviced; *(c)* they can be locally or regionally controlled and operated. These systems were co-ordinated through the required INTELSAT process and they provide services which complement standard INTELSAT services. On the whole however, the compartmentalisation of the different national, regional, public and private markets is creating a potential for challenge and intensifying competition[25].

For more specific uses, the service is presently controlled entirely by intergovernmental bodies. This is the case for maritime telecommunications, where the international

INMARSAT network is providing satellite transmission (1/5 GHZ, L Band) of data, telephone and telex to ships with suitable terminals (over 2 800 by the end of 1984). The organisation's needs are met by some INTELSAT V transponders, ESA's MARECS (1 new MARECS was launched in 1984), and to a lesser extent by the MARISAT satellites. To cover the projected supplemental demand after 1988, a request for proposals was released by INMARSAT in the summer of 1983. There will probably be a need for 6 to 8 new satellites during the next two years, and six more in 1991.

Although they are not managed by international organisations, some other international experimental systems exist, such as the search and rescue SARSAT-COSPAS system[26] which uses two Soviet COSPAS satellites and two US TIROS satellites. The US satellites will include the French ARGOS system as well as Canadian and UK components. Space systems for navigation services such as the US TRANSIT system are available for non-military users worldwide.

US domestic market

The American domestic market, which recently overtook the INTELSAT market in size, is now the largest in the world. It is also the most vigorous with a fourth-fold growth expected between 1980 and 1985. At the end of 1983, 19 civilian and private satellites were in service, all manufactured by RCA and Hughes Aircraft. Four belonged to RCA which retained the use of them, four to Western Union, and four to Comsat. SBS had three, and Hughes two for its GALAXY system. In addition, ATT was operating TELSTAR and Alascom Inc., the AURORA I Satellite. With the continuation of the existing deregulation policy, especially the dismantling of ATT, competition has developed quickly and the second generation of telecommunications satellites is gradually being put into orbit. About 16 companies have filed applications at the FCC for the launching of more than 50 satellites during the next few years (see Table 5). Half of these applications at least will expand existing net capacity.

To encourage demand and improve profitability, satellite suppliers have proposed new commercialisation procedures. These include purchasing one or several transponders in orbit[27], shared ownership and full time or temporary leasing[28]. With the development of these new procedures, the distinction between firms supplying, carrying or leasing capacities has become increasingly hazy. Various major users, such as American Satellite Corporation, are developing their own systems which will position them firmly within the market. Concurrently, manufacturers are starting to compete downstream. For example, Hughes is introducing its GALAXY System (C Band) and Ford is completing its FORD-SAT system (see Table 5).

In spite of the above developments and consequent price reductions, over-capacity in orbit is growing. A December 1984 FCC investigation showed that approximately 42 per cent of transponders were not used whereas two years before the figure was 33 per cent. Any strengthening of trends in supply therefore depends on how much excess capacity will be used to provide new services.

In this regard, significant growth is expected in data transmission via satellites and in teleconferencing (in the US, teleconferences should increase from 2.5 million to 3.3 million in 1986/87 according to a study of the ELRA Group). The large computer, telecommunication, and office systems companies are showing growing interest in this type of link (e.g. IBM's share of SBS)[29]. However, perhaps due to the large recession in the early 80s, SBS's transponders[30] (the main system with SATCOM) are still mainly used for voice-only transmissions – competing directly with ATT – and television programmes.

Table 5
US domestic carriers

Companies	Existing systems[1]	In course of implementation[1]	Extension requested	Band
Western-Union	Westar 1974 (5)	1985 (1)	1988 (10) / 1988 (3)	C / Ku
RCA	Satcom 1975 (5)	1986 (1) / 1985 (3)		C / Ku
Comsat	Comstar 1976 (4)		1988 (2)	C / Ku
IBM	SBS 1980 (4)	1986 (2)	1987 (4)	Ku
Hughes	Galaxy 1983 (3)		1986 (1) / 1987 (3)	C / Ku
ATT	Telstar 3 1983 (2)	1985 (1)	1988 (1)	C
GTE Spacenet	Spacenet 1984 (2)	1985 (1)	1986 (1)	C/Ku
	G Star	1985 (2)	1987 (1)	Ku
Am. Satellite	ASC 1985 (2)		1987 (2) / 1985 (2)	C/Ku / Ku
Advanced Business Com.		1986 (2)		Ku
Rainbow Satellite		1987 (2)	1988 (2)	Ku
US Satellite System		1986 (2)	1989 (2)	Ku
Ford Satellite Services		1987 (3)		C/Ku
National Exchange			1987 (2) / 1987 (6)	C / Ku
Alascom			1989 (2)	C
Columbia Communication			1988 (2)	C/Ku
Digital Telesat Inc.			1988 (2)	C/Ku
Equatorial Communication Serv.			1987 (2)	C
Federal Express			1988 (2)	Ku
Martin Marietta			1989 (2)	Ku
Total	(27)	(20)	(52)	

1. Date of launch of first satellite and planned number of satellites (excluding reserve satellites on ground).
Source: SEST/Euroconsult 1984-85.

Strong growth potential exists for direct broadcasting satellites (DBS) but the future is surrounded by uncertainties. Highly powered satellites with fewer channels are specifically designed for a DBS segment[31]. Accordingly, these specifications make the market entry costs high. For instance, STC, a subsidiary of Comsat – the company the most involved in this business[32] – had to reduce the scope of its initial project (a system including 4 satellites plus a spare one and ground stations for a total spending estimated at more than $800 millions). The company now intends to buy two high powered satellite (cost $115 millions) to have its service operating in 1986. Market forecasts indicate that about 10 DBS satellites will be placed in orbit before 1990, which represent at least a $1 billion investment in equipment for the space segment. The market by the end of the century is expected to reach 20 to 30 million subscribers, but this may be optimistic.

Introduction of DBS may however slow down because of demand saturation, competition by cables (about 32.3 million subscribers in 1984 according to a National Cable Television Association estimate) especially fibercables, and the growing emergence of low powered DBS systems[33]. The latter systems have an advantage as they can use more television channels on cheaper traditional telecommunications satellites (about $30 to $40 million per satellite against $60 to $70 million for high powered ones, without accounting for the differences in terms of launching costs). These lower-powered systems require heavier investment on the ground (e.g. antennae with a diameter exceeding 2.5 metres at a cost of $2 500 to $3 000), but the space segment can then be connected with cable networks.

Though of lesser importance, links with mobile users may represent a potential growth market. To service this market by the end of the decade, Canada and the United States are studying a joint project (MSAT) for automobile and truck communications. This project will include in-orbit hardware costing about $400 million and many remote area ground stations (out of the reach of cellular radio systems). Telesat will run the Canadian part of the service. Several US companies including Skylink and Mobilsat have filed with the FCC for a similar service. Approximately 100 000 vehicles could be connected by the beginning of the 90s at a cost of $2 000 per unit.

Other domestic and regional systems

Satellite orders for Western Europe and Japan - as distinguished from the US - continued to be in the early 80s either from public agencies (PTTs, national space agencies) or for export (via ESA or third markets). They were for sophisticated, general purpose systems. As regards Europe, at a national level, France has made the most appreciable commitments. Its TELECOM 1 satellite (which was launched in the Summer of 1984) was built for the Direction Générale des Télécommunications to primarily provide telephone channels and television broadcasting, but also for intra-corporate links. Also the United Kingdom has two SKYNET defence communications satellites and is developing the UNISAT system for DBS and domestic communications[34] (3 satellites). The German DFS/KOPERNIKUS telecommunications television distribution satellite should be in orbit by mid-1987 and two French/German television satellites, TDF-1 and TVSAT, will be launched in 1986. Japan's first operational domestic telecommunications satellites, CS2A and CS2B, were put into orbit in 1983[35].

In Canada, the domestic satellite communications system is owned and managed by Telesat. This corporation represents an important customer for satellites. Between its creation in 1969 and 1985, Telesat will have procured, launched (by NASA) and operated 9 ANIK commercial satellites (3 ANIK A, 1 ANIK B, 3 ANIK C, 2 ANIK D) to provide a variety of continuing telecommunications services to Canadians.

Although the concept of European satellite telecommunications originated in 1970, the EUTELSAT Organisation was only founded in 1977 (as an interim organisation comprising 17 countries). Two missions were initially assigned to this new institution:

- creation and operation of a European regional satellite telecommunications system for fixed service applications;
- provision of new means of linking the radio and TV broadcasting organisation members of EBU (European Broadcasting Union) for Eurovision programmes.

For this purpose the ECS programme was developed, following the experimental OTS project. It has recently been enlarged and the creation of a satellite multiservice system (SMS) has been selected in order to offer a wide range of telecommunications services requiring only relatively simple earth station facilities. 20 are planned under the ECS multiservice operation, which will start soon. This new service will use part of ECS's space segment in conjunction with part of France's TELECOM 1 space system. ECS1 spacecraft (launched in 1983) will also be utilised for a project mixing low power DBS with cables networks in the UK, Switzerland, Belgium, the Netherlands and the Scandinavian countries (this project is handled by the SAT-TV group). In addition to ECS2 placed into orbit in 1984, and ECS3 planned for 1985, five new satellites should be built in the coming years to cover EUTELSAT needs.

Several factors account for the differences between the European oligopsony and the US market structure, such as:

- the influence of institutional demands, traditional in certain countries;
- the early lead of North American space programmes. European firms have not yet taken full advantage of the learning curve and consequently cannot fully control costs, and their penetration of high-risk markets is still subject to the existence of public-sector demand;
- the relatively recent creation of the European Space Agency;
- the heterogeneity of Europe which deprives the industry of adequate economies of scale (e.g. different languages serve as an impediment to the extension of television networks, slow rate of harmonization for technical standards (see Chapter VII section 4), etc.);
- the geographical structure of Europe as a whole (concentrated population, density of conventional networks).

2. SATELLITES FOR WEATHER FORECASTING AND OBSERVATION/REMOTE SENSING

Special-purpose satellites form smaller, fairly stable markets. For weather forecasting, worldwide launchings are expected to continue at their present rate of two per year. This continuity in trend should lead to regular growth for the market in specialised ground terminals.

Although there is no doubt about the value of meteorological programmes (the American TIROS, NOAA and GOES, the European METEOSAT and the Japanese GMS), the transfer of responsibility to weather forecasting agencies has been gradual. In Europe this responsibility has been invested in the EUMETSAT organisation, which ordered in 1984 three meteorological satellites using the technical support of ESA.

Space meteorology is an area of large international co-operation. For example, over 120 countries receive direct broadcasts from the US weather satellites and even more countries receive satellite data through WMO's global telecommunications system. These data are all available free of charge.

In remote sensing[36], the United States is the only country providing data and informations, which it derives from its LANDSAT satellites. The information is nevertheless received and distributed by various national and international bodies, e.g. ESA EARTHNET service. Although there is an appreciable potential market for customers such as mining and oil companies, agriculturalists and government departments managing hydrological

26

resources, remote sensing does not yet cover its operating costs (approximately $30 million in 1984, while revenues amounted to some $10 million). Hence, further fee adjustments are likely.

Nevertheless, with the launch of the French satellite SPOT[37] scheduled for 1985, at a capital cost of nearly FF 3 billion, and the formation of a commercial entity, i.e. the marketing company Spot-image and its distribution network, keener competition may be expected. Other satellites such as ERS for ESA, MOS for NASDA in Japan and RADARSAT in Canada will also be placed in orbit during the coming years. The overall worldwide market is expected to amount to about $1 billion by 1990.

3.   EARTH STATIONS

A satellite system requires not only satellites (space segment), but also numerous earth stations (ground segment). These earth stations or terminals vary according to their purpose including: international and/or domestic communications, remote sensing, search and rescue, metereology, and TT&C (telemetry, tracking and control). With the exception of the TT&C stations, which control the satellite and are normally purchased in conjunction with the first satellite acquisition, the earth stations are generally purchased separately from the satellites, although they must be technically compatible.

Terminals vary in size and complexity. Some are large infrastructures with 30 metre steerable antennae weighing many tonnes and costing millions of dollars, such as INTELSAT Standard A stations which can carry heavy intercontinental telephone, data and video traffic. Others are small equipment such as 0.9 metre Television Receiver Only (TVRO) antennae costing less than $700 for home TV reception. As the industry develops, there is a tendency for the manufacturers to specialise in one or a few product lines (e.g. INTELSAT's standard stations, TVROs, low-cost telephony terminals, or remote sensing terminals), although there are some firms such as Scientific Atlanta which produce a wide range of related products (e.g. ranges of TVROs).

Earth stations represent a highly competitive market which is far less compartmentalised than that of satellites. The competition is based mainly on price and delivery and not on technical performance because suppliers must guarantee a performance which meets the technical standards of INTELSAT, INMARSAT, etc. For the very large Standard A, B and C stations, Japanese, North American[38] and European producers all have a continuing share in the market and compete vigorously internationally[39]. This is also the case for ship terminals. Other market segments, especially for small TVRO dish services, have yet to emerge significantly outside North America where there are already more than 40 manufacturers. In Europe, however, the influence of public-sector demand benefits national producers, although the recent reductions in station size should encourage greater competition.

The overall demand for the ground station segment, i.e. about $10 billion during the 80s[40], has been expected to grow rapidly (by 35 per cent annually between 1980 and 1990 for guidance stations). In the United States, consumption of commercial and private earth stations should increase from $ 198 million in 1983 to 425 million in 1987 according to Electronics Magazine.

Generally ground segment related expenditures exceed satellite spendings. This is because the equipment is expensive and the incoming signal is so weak that highly sensitive equipment is required. Thus, FF 0.5 billion are invested in building the 300 stations used for the French TELECOM 1 project bringing the total cost to FF 1.5 billion. For the INTELSAT network (573 stations in service by the beginning of 1982), procurement for the earth segment is estimated to have cost £100 million/year over the last 10 years.

In fact, the development and size of these markets vary considerably depending on the kind of terrestrial system concerned. Thus the recent downturn in the market for larger antennae due to the saturation of existing demand and the introduction of relay satellites will probably be offset by vigorous growth in the demand for small dishes used for DBS reception[41]. By 1990, according to some analysts[42], demand for the latter in the United States could amount to $2 billion per year (at least 5 million homes to be equipped), and to $0.3 billion (for a minimum of 1 million direct TV receivers) in Japan. The industry as a whole will probably be further invigorated by exports to the Third World (the developing countries' market for earth and space segments combined is estimated from $3 to $3.6 billion between 1982 and 1987). Lastly, the number of ships equipped with INMARSAT terminals could reach 10 000 by 1990 (at a unit price of $25 000).

Chapter IV

PRICES, COSTS AND COMPETITION

1. SPACE SEGMENT

The rise in satellite prices has been fairly moderate since the beginning of their commercialisation. This is due to the striking rise in transmission capacities of telephone circuits and TV channels[43] for telecommunications satellites and increases in the precision and performance of remote sensing. These developments have resulted in substantially lower costs, making both satellite equipment and the services they provide, increasingly competitive[44] (see Graph 2 and Annex I.8).

This trend can be related to satellite weight gains but it also owes much to changing technology in telecommunications, such as alterations in microprocessor technology and electronic circuits, and in multiplexing techniques (e.g. Time Division Multiple Access, TDMA), creating higher yields and greater capacity. The use of higher frequencies (4/6 and 12/14 Ghz presently – experimenting with 20/40 GHz) means that new services can be offered as a way around the existing congestion on lower-frequency bands. Satellite lifetimes have been increasing[45] – up to ten years for the future INTELSAT VI series – especially because of the progress made in amplifying materials. They are now more able to withstand widely fluctuating temperatures in space. Lighter, more efficient fuel and lower energy consumption ratios have also played a key role in reducing costs.

Although public-sector markets continue to account for a substantial proportion of world demand, the recent increased competition among country producers also plays an important part in curbing equipment costs. To be awarded a contract in a public-sector or private-sector call for tender, firms have been induced to find ways of improving cost efficiency, thereby designing less expensive space vehicle configurations (e.g. integrated propulsion systems) and to use more or less standardized platforms. However, the most notable economies have been achieved in the manufacturing of cells and components where competition has been most vigorous. It extends across frontiers with the development of international programmes (as mentioned above), but overall imports are still limited. Distinct imbalances in the volumes of the US market and its related scale effects as against those of Japan and Europe[46] help to maintain appreciable cost differentials – 15 to 25 per cent according to certain sources – between producers in different areas. However, when production runs are more extensive, as in sub-contracts for new series of satellites for INTELSAT, non US-space companies are able to demonstrate increasing competitiveness.

2. GROUND SEGMENT

Due to the advancements in satellite technology, some of the earlier technical requirements for the construction and operation of ground stations have been reduced. Previously, all terrestrial stations required tracking antennae to keep in touch with

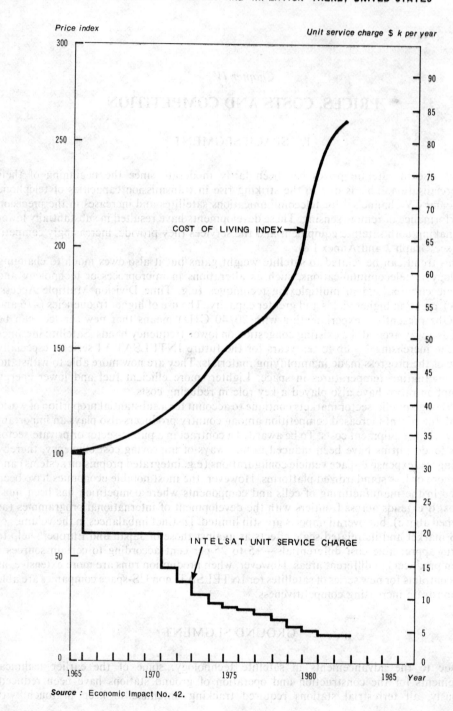

Graph 2 INTELSAT CHARGES AND INFLATION TREND, UNITED STATES

Price index

Unit service charge $ k per year

COST OF LIVING INDEX ➜

INTELSAT UNIT SERVICE CHARGE

1965   1970   1975   1980   1985   Year

Source : Economic Impact No. 42.

non-geosynchronous satellites and, when they were transmitted over very wide areas, the signals were usually so weak that they required more sensitive receiver aerials and therefore larger diameters. With increasing transmission power, progress in guidance (stabilization triaxially or by rotation) and more accurate beaming, earth station investment costs have been considerably reduced. Furthermore, amplifier technology innovations (transistorised parametric amplifiers) have helped bring down the cost of earth station equipments[47]. The new amplifiers can work in ambiant temperatures and are more reliable and easier to maintain. With the cost of amplification, the cost of reception has declined. Satellites have become more specialised and complex antennae have become smaller and generally less expensive (e.g. average TVRO dish price declined in the US from $25 000 in 1978 to less than $5 000 in 1983). Also the possibilities created by using relay satellites[48] could lead to diminishing earth terminal investment costs.

3.   LAUNCHING AND RELATED SERVICES

For the satellite owner or operator, the cost of putting it into orbit represents roughly 20 to 25 per cent of the total investment (i.e. satellite acquisition, launching and control station cost). This relatively high cost can be attributed, *inter alia,* to:

-   the heavy per flight costs in the early years of new space systems;
-   the complexity of assembling the different parts of the rocket thus resulting in relatively long production cycles and the need for rigorous scheduling;
-   the volumes and prices of fuel required (several million dollars for liquid hydrogen, liquid oxygen, hydrazine, N204, etc.);
-   the necessary investment in technical services (qualification tests and launch follow-up), involving the permanent or temporary employment of highly skilled staff;
-   the present tendency towards more powerful rockets, as satellites become more sophisticated and heavier;
-   the risk factor which is to a certain extent measured by the cost of insurance cover[49].

Up to now, considerable government involvement in expenditure in manufacturing countries (covering research and infrastructure development costs) has meant that the relationship between customer launch charges and real costs is not easy to establish. At the same time, some civil rockets were derived from a missile (see Table 6) and a considerable proportion of the expenditures in building the prototype and developing early models will thus have been covered by defence budgets. In other cases, such as the Space Shuttle, certain space transport systems are used for both civilian and military missions[50] which makes the price setting process complex.

In general, launch prices are based on an estimation of a cluster of basic factors such as the period of time necessary to write off the initial investment, the projected number of flights during this period which depends on demand expectations and the projected cost and desired rate of return on the investment. However, in the space business, not all fixed costs[51] are recouped due to the specific nature of this activity and the policy aspects underlying government decisions to establish the capacity to operate in space. In the United States, for example, STS is considered a national resource which would be maintained for public purposes even if foreign and commercial reimbursable missions were not scheduled. Therefore, R & D spending and in general non-recurring costs are not included in the cost base for the pricing of STS launching service.

Table 6

Characteristics of the main operational civil launchers

Country of origin	Name[1]	Main Type	Manufacturer	Technical features	No. of successful flights (february 85)	Basic payload weight (Tons) on transfer orbit
United States	STS (Shuttle) Present fleet: 4 Orbiters	R	Rockwell International for the STS; Morton Thiokol subcontractor for the solid rocket boosters; Martin Marietta for the external tank; Rocketdyne for the orbiter main engines		14	PAMD:1.2 PAMA:2 IUS:4.5 CENTAUR (86):13
	DELTA	NR	Mc Donnell Douglas	Derives from Thor missile	158	1.25 (Delta 3920)
	ATLAS	NR	General Dynamics	Derives from a missile	185[2] (60) with Centaur upper stage	2.4 (G/Centaur)
	TITAN	NR	Martin Marietta	Derives from a missile	137[2]	4 (34D+IUS)
Europe	ARIANE	NR	Arianespace	Derives from French rockets Véronique and Coralie	10	2.4 (Ariane 3)
Japan	N-II	NR	Mitsubishi heavy industries	US licence, derives from Thor Delta		0.4 (Geostationary orbit)
URSS	PROTON	NR		Korolev rocket	81[2]	4 (Geostationary orbit)

R = recoverable; NR = non-recoverable.
1. The name of of the rocket is the name of the first stage; upper stages can vary.
2. 1957-1982.
Sources: Euroconsult/Espace/83 and OECD Secretariat.

32

In view of the present international debate on launching prices, it is difficult to expand on this. In brief the price for chartering the entire Shuttle cargo bay has been fixed by NASA at $38 million (FY 82) for the period FY 82-85 and $71 million (FY 82) between FY 1986-88. Additional information on ELV prices is available in Graph 3.

With growing demand for space missions leading to numerous space flights and probable increases in operational efficiency, launch cost trends could slow down or even reverse. This trend, together with the price catching up process, may lead relatively rapidly to more practically free market conditions. The first indications are given below:

i)   several companies in the United States are now attempting to establish commercial launch services, with some even proposing to develop their own rockets (e.g. TRUAX Engineering and its EXCALIBUR rocket, Space Service Incorporated and its CONESTOGA rocket – this last rocket has been successfully tested in a

Graph 3   **LAUNCH PRICE PROGRESSION FOR DELTA CLASS SATELLITES**[1]

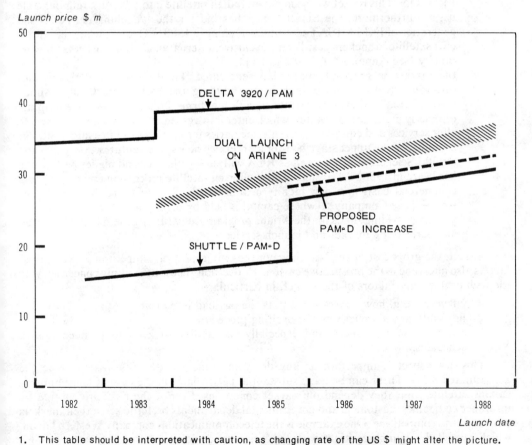

1. This table should be interpreted with caution, as changing rate of the US $ might alter the picture.

Source : Trends in Satellite Communication. A.D. Wheelon, World Telecommunication Forum, Geneva, October 1983.

suborbital flight). As the Shuttle flight rate begins to build towards its full intended rate, NASA is phasing out its expendable launch vehicles (ELVs) and transferring these systems to the private sector. Use of government launch facilities will be offered for a fee. During this transition period, a number of Shuttle and DELTA compatible satellites have been launched on the DELTA ELV, as have several communications satellites on the ATLAS CENTAUR. In addition several scientific missions requiring polar orbits have been launched on the DELTA ELV, while NASA awaits completion of the Shuttle polar launch site at Vandenberg Airforce Base, California. The last NASA ELVs are scheduled to be phased out by 1986;

ii) the Western launching monopoly of the United States, and NASA in particular, ended in 1981 with the qualification of the European launcher ARIANE, especially for injection into geostationary orbits. Since 1979 Japan has been launching its satellites with its own launch vehicles. Japan plans to develop a new launch vehicle H-II for heavier satellites in the early 90s. In addition the USSR is showing increasing interest in the commercialisation on world markets of its launcher PROTON. This rocket will soon put an Indian satellite into orbit on a reimbursable basis. Furthermore, the Soviet Union has bid for the launching of the second generation of INMARSAT satellites beginning in 1988[52]. Overall, while the club of civil satellite launchers is still very exclusive, international competition is growing rapidly (see Annexes I.9, I.10 and I.11);

iii) the market is gradually becoming segmented. For several years the various functions involved with a launch, i.e. building the rocket, marketing, launch services, etc., have been performed by different firms. New companies are forming, especially in the United States, which enter into agreements with manufacturers to offer services and equipments as intermediaries for putting satellites into orbit. For example, on the upper stage booster market two new space ventures, Astrotech and Orbital Sciences Corp., as well as RCA will supply cheaper and higher performing hardware in 1986[53]. Astrotech is also offering satellite processing and testing on a commercial basis in a new facility near Kennedy Space Centre. In Europe, the Arianespace Company[54] (whose capital is held by private and public interests) is already providing funds for the Ariane program, constructing the ARIANE rocket, and performing market and launch services.

In fact, the gross cost of the launch is not the only aspect of competition. "Non-price" factors also intervene in the prospective customer's decision. These are mainly concerned with the technical quality factors of the rocket, in particular:

– efficiency, e.g. how accurately it puts the payload into orbit;
– flexibility and simplicity of the orbiting procedure;
– reliability of equipment and especially the ability to keep to planned launch schedules.

This last aspect is important as any delay in launching involves losses due to the accumulated delay. These can be quite substantial particularly in the case of telecommunications satellites, but they depend on market conditions. Even if the risks are covered by insurance companies, customers and governmental departments prefer to serve their markets on time due to competition. One example is the telecommunications company Western Union, which cancelled a launch contract with the Arianespace company to take advantage of a possible option on the Space Shuttle three months before the date initially proposed and despite the penalisation cost for breach of contract.

Chapter V

SPACE PRODUCTION AND PUBLIC BUDGETS

Space production began as a natural extension of the aeronautics industry: the flight range of craft from the stratosphere into space was merely extended. Apart from its scientific origin and the military interests which governed its development, space production retains many features of the aircraft industry including the cyclical nature of the activity, rapid innovation rates, substantial R&D investment – about 5 per cent of sales in the United States – using large amounts of public funds[55] and a high proportion of qualified staff, technicians and engineers. This linkage is reflected in the organisation charts of major aircraft companies most of which now include ballistic and space divisions.

The space sector's share of world aerospace sales, excluding planned economy countries, rose quickly from almost zero in 1955 to 25 per cent in 1965 under the influence of NASA's space programmes (see Chapter II). Due to United States cutbacks during the 1970s, which were only partly compensated by the boom in the European and Japanese programmes, this share then fell to 12 per cent in 1980 or approximately 7.6 billion dollars. Current developments, notably for satellites with the rapid growth in communication needs, suggest that this trend has bottomed out, especially since the long recession following the second oil shock has led to stagnation in air transport.

Space products development and production have so far depended to a large extent on public investments. However, that picture is changing, as privately owned and operated satellite systems become more common and the launching of those satellites becomes competitive. The industry is similar to aircraft manufacturing as it is now developing and the service side of the industry is similar to the aviation transportation sector. Whether or not the industry and the resultant services are publicly-funded depends on the economy in the various countries.

In most countries assistance to space production consists of research contracts, procurement of prototypes or experimental spacecraft, R&D financing and various appropriations for technological developments. This may mean placing the testing or equipment facilities of public laboratories at the disposal of enterprises, who can check and approve the equipment. For example, with launchers, the space agencies take on many R&D expenditures and staff costs. They pay all fixed costs connected with infrastructures (site, equipment and overall maintenance). Full coverage is also provided for scientific payloads (e.g. space observatories, astronomical observation satellites, orbiting scientific laboratories). On the other hand, application contracts such as those for space telecommunications do not necessarily receive substantial R&D support, even in absolute terms, especially in the United States.

Therefore, it is not surprising that the world distribution of space activities fairly closely reflects the world distribution of space budgets (see Tables 7 and 8 below).

World production is still considerably dependent on the United States (80 per cent of the world total in 1980), although its share has been falling for the past few years due to budgetary policy and the internationalisation of markets. Moreover, the success of Canadian and British enterprises in export sales should be noted. Table 7 shows that the aggregate turnovers for Canadian and British industry considerably exceed total space expenditures (these are, in fact, the only examples of this situation).

Table 7

**World distribution of total space budgets and size of space industries
(in million US dollars) for the main countries in 1983**

	Total space budget	Turnover of space industry*
United States	15 048	5 900
France	404	380
Germany	299	250
Japan[1]	477	400
Canada[2]	109	150
United Kingdom[3]	98	180
Italy	109	120
India	98	40
Netherlands	53	25
Sweden	40	20
Belgium	34	20

* Including all types of space hardware (launchers, satellites and equipments) excluding services and ground segments. Calculations are based on public administration data (e.g. United States Department of Commerce) and/or industrial associations data. Because of differences in national definitions, difficulties in aggregating telecommunication and aeronautic data and in classifying equipments according to their space or non-space utilisation, these figures should be considered as estimates with an order of magnitude of space activities. Therefore one should be cautious in making international comparisons.

1. Turnover for the Japanese space industry as a whole (i.e. including earth stations) is for 1978: 491.2, 1979: 430.1, 1980: 566.7, 1981: 507.3 in United States dollar (on the basis of $1=Y240).
2. Turnover for the Canadian industry as a whole is in C$: 1978: 111, 1979: 138, 1980: 120, 1981: 123, 1982: 196, 1983: 276, 1984: 313 (estimation).
3. Sales of space products in 1978-1981 by the United Kingdom space industry as a whole as published by the Society of British Aerospace Companies are: 1978: 34.7, 1979: 85, 1980: 112.5, 1981: 123.7 (in £m).
Source: SEST/Euroconsult, 1984-85.

Table 8

Budget and staff of main civilian space agencies in 1981

Space related activities	NASA US	ESA Europe	CNES France	NASDA Japan	ISRO India
Budget (in $ millions)	5 030	700	467[1]	449	68
Staff	22 100	1 400	1 160	894	7 021[2]

1. Of which $180 million allocated to ESA.
2. This figure cannot compare with the others, because of differences in definitions.
Source: Aeromag.

Under various government policies, the priority given to space activities varies from one country to another. Again the United States is in the lead (see Table 9 below).

Table 9

Space expenditure per capita in 1983 (in ECU)

	A. Space exp. per capita	B. GDP per capita	A/B × 10⁴
United States	74.1	14 985	50
Belgium	3.1	9 500	3
Denmark	2.5	12 500	2
France	9.9	10 860	9
Germany	5.7	12 270	4.5
Italy	2.7	7 200	3.7
Netherlands	2.9	10 660	2.7
Spain	0.6	4 725	1.3
Sweden	9.1	12 635	7
Switzerland	2.5	17 360	1.5
United Kingdom	1.8	9 200	2
Ireland	0.4	5 780	0.6
Austria	0.26	10 183	0.2
Norway	0.8	15 320	0.5

Source: Eurospace.

NATIONAL SITUATIONS CONCERNING PUBLIC EXPENDITURE ON SPACE ACTIVITIES

a) **United States**

In constant dollars, NASA's budget remained unchanged in 1982 ($5.8 billion) from the previous year, but increased in 1983 by about 8 per cent. It reached $7.2 billion in FY 1984. With the share of expenditure on the Space Transportation System (STS), including the Shuttle (approximately 55 per cent for total budget, 63 per cent of R&D expenditure), the amounts spent on application contracts are falling in relative terms, for telecommunications ($30 million) and also for weather forecasting and remote sensing. For the first time, the space station will be funded in the 1985 budget to the amount of $150 million. For R&D programmes as a whole, the Administration is aiming at reorganisation to promote basic research which will lead to a decrease in applied research. The latter is gradually being left to the private sector, which is considered more capable of selecting suitable techniques for the market. NASA programmes may be affected by this policy in 1985, in particular research on advanced telecommunications satellites and advanced propulsion systems.

There is a certain shift towards the Department of Defense, which is presently the main source of funds for telecommunications with 7 satellites constructed in 1983; it provided already $800 million in 1982, of which 300 million were for R&D. For space activities overall, according to NASA's estimates, an even more marked shift is towards the private sector about 32 per cent of space expenditure in 1984 (as against 18.5 per cent in 1975), the shares of DOD and NASA being 38 and 30 per cent respectively in the same year. In R&D, however, growth in expenditure by firms was not sufficient to compensate shrinking federal funds, thus causing an overall decline.

b) **Canada**

The Canadian space programme was initially motivated in the early 1960s by the need to improve national communications services, particularly for the isolated northern populations. With the realisation that communications satellites effectively met these needs, the authorities concentrated on assisting the development of the domestic industry's capability to supply the necessary equipments and services. Co-operative projects with the United States and European countries and industry supplemented the extensive R&D carried out by the Canadian Government and industry. These efforts culminated in the development of Spar Aerospace's capability to prime contract the ANIK Ds and subsequently BRASILSAT, as well as the RMS – Canadarm for NASA. At the same time, other Canadian companies developed expertise in specific technologies and product areas (e.g. Comdev in satellite multiplexers, SED Systems in TT&C stations and satellite receivers, Canadian Astronautics Ltd. in SARSAT local user terminals, Microtel in Spacetel telephony terminals).

The Government's current space programme includes major participation in ESA's OLYMPUS programme (C$100 million), research on MSAT, a dedicated satellite for mobile communications (with launch anticipated around 1988) to be carried out jointly with NASA, the development of the RADARSAT remote sensing satellite (C$520 million of which Canada is providing 300 million) and earth station technologies, also with foreign partners, as well as satellite sub-systems and the possibility of DBS satellites. Space programme expenditures planned for 1985/86 in the Interim Space plan will be approximately C$195 millions (including 8.8 millions for the space station studies), i.e. a 30 per cent increase on 84/85 expenditures.

Canadian companies' practical applications, proven products and expertise have contributed to their success in export markets. In addition to communications satellite technologies, Canadian companies have relative leads in certain aspects of remote sensing, search and rescue and robotics applications (RMS).

c) **Europe (ESA)**

ESA had in 1983 a budget of $705 million, while total space expenditures in Europe were estimated at more than 1 billion.

The ESA efforts since its creation (1975) have mainly been centred on the development of:

- the ARIANE-I launcher, now operational (roughly estimated at $835 million at the 1983 price level, on the basis of $1 = 1.15 European currency unit);
- the SPACELAB manned laboratory, first flown on the US Shuttle in November 1983 ($660 million at 1983 price level);
- telecommunications satellites (OTS, ECS for EUTELSAT, MARECS 1 and 2 for INMARSAT, OLYMPUS) and other application satellites (METEOSAT, etc.);
- scientific satellites (GIOTTO and HIPPARCOS were the latest ones undertaken) and a number of joint scientific projects in co-operation with NASA (ISEE, Space Telescope and ISPM).

d) **France**

In France, the expenditure breakdown reflects the importance of multilateral co-operation within ESA (more than 35 per cent of the 1983 CNES budget). In this context, emphasis is placed on the development of the ARIANE programme (where the French contribution exceeds 60 per cent). France also participates in all types of commercial space activities, either through specific efforts such as the TELECOM 1 satellite programme, or through bilateral co-operation, for example the TDF/TVSat direct broadcasting satellite (with Germany). In addition, CNES has developed an ambitious remote sensing programme (funding FF 920 millions in 1984) largely devoted to the SPOT project. A major budgetary appropriation was granted for R&D, which accounts for 3.7 per cent of CNES total budget. In addition to satellite systems and earth observation techniques, the main programmes relate to inter-satellite communications, the development of spacecraft, the evaluation of robots for orbit stations, data relaying systems, the space aircraft HERMES and the next 80-100 tons thrust cryogenic motor HM 60. Public appropriations for CNES were approximately FF 4 billion in 1984.

e) Germany

Whereas in France national projects account for nearly two third of the space agency's budget, in the Federal Republic of Germany they total merely 50 per cent of the space budget for the Ministry of Research and Technology (BMFT). National projects cover extraterrestrial research, remote sensing, telecommunications and material science. The scientific X-ray satellite ROSAT is carried out on a trilateral basis with the United States and Great Britain. The DBS satellite TVSAT is under development jointly with France. The R&D work on telecommunications supported by BMFT so far has enabled industry to take over the development and manufacturing of the telecommunications satellite KOPERNIKUS ordered by the Deutsche Bundespost. Based on a large engagement in the European space laboratory project SPACELAB (see Annex I.12), the utilisation of the Shuttle/Spacelab system is emphasized in the areas of materials science, remote sensing and extraterrestrial research. Government total expenditures for space technology are planned to increase from DM 680.3 in 1984 to DM 736.3 in the 1985 budget.

f) United Kingdom

In the United Kingdom, two facts are striking: the relatively small size of the space budget and its emphasis on multilateral co-operation (80 per cent goes to ESA) plus the strong concentration of activities on telecommunications (all the prime contractors for ESA satellites in this field are British) and other application satellites. In this context, the most important United Kingdom contribution adresses to the MARECS maritime European communication satellites (about 60 per cent). The United Kingdom is also developing a direct broadcasting TV satellite UNISAT and the SKYNET defence communications system.

g) Italy

In Italy, for a long time space research was carried out by the universities and the armed forces, but a reorganisation took place when Italy joined ESA. Most funds are channelled towards the future national satellite ITALSAT and within ESA, to the OLYMPUS satellite. Telecommunications receive priority, although a few launcher activities are being maintained, for example, the construction of booster rockets for ARIANE by SNIA-Viscosa and the development of an IRIS upper stage by Aeritalia and BPD Difesa e Spasio. This upper stage will be tested on the Space Shuttle in 1987. Italy is also a major participant in the SPACELAB programme and is contributing through a co-operative project with the United States to a Tethered Satellite System.

h)   Sweden

The principal objective of the Swedish space programme is to encourage the development and diffusion of advanced technology within Swedish industry. In 1979, Parliament adopted this new policy and decided to triple the level of expenditure. The Swedish Board for Space Activities, under the Ministry of Industry, is the responsible governmental agency and the state-owned Swedish Space Corporation performs executive functions for the Board. Sweden's first satellite VIKING, scheduled for launch with SPOT in 1985, is a scientific satellite, which will study the magnetosphere at high altitude over the North Pole. Saab-Scania is the main contractor while Boeing is responsible for the platform. The second satellite, TELE-X, scheduled for launch in 1986, is a multi-mission experimental/preoperational satellite with data and video communications and direct TV broadcasting for Sweden, Norway and Finland. The project is jointly governed by Norway and Sweden with Finland maintaining observer status. Norway owns 15 per cent of the consortium's shares and is financing 15 per cent of the total project costs. Aerospatiale is the prime contractor. Sweden is involved in bilateral industrial co-operation projects (e.g., with France for the SPOT satellite) and is making a major effort in the remote sensing field.

i)   Japan

In Japan, the total space activities' budget in FY 1984 was 113 billion yen ($470 million). However the trend of government expenditures for space activities in the 80s does not show the kind of growth that was taking place in the middle 70s.

There are two space programmes – applications and science – conducted by two different organisations. NASDA, which is in charge of the space applications programme, spent 84 billion yen ($350 million) for its own programme in FY 1984 and ISAS (Institute of Space and Astronautical Science) had a 16 billion yen ($66 million) budget for space science in FY 1984. The current plans for the NASDA budget show that about 33 per cent is allocated for satellite developments (see Annex I.13).

The political will to attain possession of adequate technological capabilities for launch vehicles, satellites and other space-related areas has secured continued support for the development of these objectives. The goal for Japan is to establish its own technological capabilities, not only so that it can carry out its space programme at its own discretion, but also to enable it to play an important role in international co-operation with its technology.

j)   Other countries

In the OECD area, apart from the countries already mentioned, space activity is confined to Australia and a few smaller European countries. ESA has succeeded in getting the participation of the latter group in the agency's programmes. Usually, the participation does not exceed 5 per cent of the Agency's total budget. Among this group, the Netherlands, which is mostly geared towards scientific satellites, is the only country to have a national programme. Belgium contributes up to 5 per cent of the funds for the ARIANE programme and is also involved in the French SPOT programme. Swiss efforts in the space area are mainly taking place through their participation in ESA. Within this framework, their most important contributions are addressed to the ARIANE programme and to scientific projects.

40

Chapter VI

STRATEGY OF FIRMS AND THE SUPPLY STRUCTURE

Given the high cost of finished space products and their high R&D content, the space market structure mainly reflects the effects of size, the technical experience of firms and their advance in the field. As between countries, it also reflects the importance of the role played by national civilian and military programmes[56] rather than differences in organisation, management or financial risk-taking. Access to R&D, the creation of expertise, and more generally the development of firms' production capabilities are all influenced by these programmes. Furthermore, this influence is even stronger because trade is limited.

1. STRATEGIES OF COMPETITION AND/OR CO-OPERATION[57]

In the space business, co-operative efforts between firms are common because of, *inter alia,* the high cost and risk involved in making new investments and the great variety of components and sub-systems required to manufacture the final product. In more mature markets, however, firms prefer to compete independently if they have reached a certain size and technological autonomy.

In the case of the United States, the emphasis leans towards competition and market segmentation, but existing circumstances and the related company policies vary considerably. Hughes Aircraft, the leader in satellite sales, has 30 per cent of the US civilian market. Its activities encompass all segments of the market (platforms, payloads and earth stations) and

Table 10

Size and staff of United States companies (1982)

Space related activities only

Companies	Turnover $ million	Staff
HUGHES	615	7 000
RCA (ASTRO-ELECTRONICS)	300	1 400
TRW	485	4 500
FORD	400	4 500
GE[1]	350	4 500

1. 1981 figure.

it has a turnover far beyond France's entire space budget (cf. Table 10). The company aims to further strengthen its leading position as a satellite prime contractor and thus continue deriving maximum benefits from economies of scale[58]. It is doing so by diversifying its activities. Previously Hughes only manufactured satellites, but it recently established a common-carrier type network to lease transponder capacity (similar to RCA's original satellite network). This GALAXY network enables Hughes to penetrate the highly profitable services sector[59] and develop a production schedule based upon the company's internal demands. Other companies aim to increasingly specialise their product lines. This is already evident with RCA's meteorological satellites, and in the future will apply to their heavy DBS projects. General Electric and TRW, whose satellite activities depend heavily on military contracts, are concentrating on remote sensing technology. Other companies which maintain a relatively more modest share of the US market, such as Ford Aerospace, place higher priorities on co-operative projects with foreign entities.

Nevertheless, even if some major companies have the experience and capacity to produce and assemble fully integrated products such as a satellite and its load, the finished product integration is generally based upon subcontracting. This sourcing is especially widespread for launchers. Industrial teams are usually responsible for Shuttle or ELV manufacturing and launching. For example, McDonnell Douglas is the manufacturer and vehicle integrator of the DELTA rocket, whose management is being passed on to TCI. Various subcontractors for engines and fuel include Morton Thiokol, Rocketdyne, Aerojet and TRW.

With the present trend towards commercialisation, new service companies are increasingly moving into the space business (three firms offered space services in 1980, twenty-five in 1983). Because of the high start up cost and relatively small amount of venture capital available, affiliation with larger companies has been commonplace with the newer entrants. Co-operative ventures have also developed in order to attract seed money, share risks and take advantage of technological complementarity.

The latest United States' trends reflect company strategies that generally emphasize *i)*, the exploitation of technological advantages and *ii)*, the widening of product range.

i) Hughes is improving its existing hardware (especially with the new HS 383 platform, which will be launched for the first time in 1985) and increasing its research effort in the field of high frequency 20/30 GHz. One satellite will be placed into orbit in 1988 at a cost of $450 million. RCA is also emphasizing this market segment with the NASA ACTS advanced communications programme and introducing innovations such as an ion engine and antennae with a reconfigured beam pattern. TRW is seeking to utilise the experience gained with the development of the TDRS.

ii) Ford presented at the end of 1983 its new series of satellites called SUPERSAT. This series included various types of modularized satellites: multiservice, hybrid, medium and high powered DBS. The SUPERSAT family is using a basic bus and a high performance propulsion module. Simultaneously, Hughes is preparing the ground for the DBS market (with its new HS 394) and is expanding its product line by offering a cheaper satellite designed for developing countries needs (HS 399).

Smaller firms, for example Canada's Spar Aerospace[60] are sometimes able to pursue multipurpose strategies. After establishing itself in satellite subsystems (payloads) and earth stations, Spar has now developed a satellite primecontractor capability and is working on developing satellite structural (bus) capabilities (see Table 11).

42

Table 11

Size and staff of Canadian companies (1982)

Space related sales only

Companies	Turnover $C millions	Staff
SPAR	108	2 000
SED Systems	21	270
MACDONALD DETWILLER & Associates, Ltd.	15.5	250
COMDEV, Ltd.	9.6	190

In Europe, on the other hand, companies are more inclined to operate cooperatively since the market is limited and producers are smaller and more specialised (see Table 12 and Annex I.14)).

Table 12

Presentation of a few european companies (1983)

Country	Companies	Staff involved in space programmes	Estimated space turnover in country ($ millions)
France	AEROSPATIALE	1 800	140
	MATRA	1 100	160
United Kingdom	BADG	1 700	140
	MSS	700	
Germany	MBB-ERNO	2 000	190
	DORNIER	600	
Italy	SELENIA-SPAZIO	700	105

The major European space firms teamed together during the sixties into three consortia: MESH, COSMOS and STAR, (see attached list of consortia members and their role in Annexes I.15 and I.16), mainly to carry out ESRO or ESA scientific programmes. Thanks to these consortia, preferential links between a number of contractors were created, and new structures have recently evolved, which are more adapted to the penetration of the commercial space markets.

Alongside the consortia, joint-ventures have accordingly been set up like EUROSA-TELLITE and SATCOM INTERNATIONAL (see Annex I.17). Their aim is to try to use the products and technologies developed and the experience gained from ESA and national programmes during the seventies to compete for export markets.

Association with non-European manufacturers is significant. Examples are the Ford/Aérospatiale agreements for the building of the Arab League ARABSAT satellite, the INTELSAT satellites, the MSS/Ford, the BAE/Spar/Selenia and Hughes/BAE agreements to compete for the new INMARSAT satellites generation, etc.

On the international market, apart from its role in the ARABSAT Programme, the European space industry has been successful in supplying sub-systems through sub-contracting for the various INTELSAT families of satellites. The penetration of the United States market has been difficult and has primarily consisted of ARIANE launch services for some domestic telecommunications satellites.

In order to better adjust to demand and to diminish costs through subsystem standardization, European manufacturers are now attempting to set up new satellite series, which are more versatile and are for multiple uses such as DBS, telecommunications and mobile links. For this purpose, Aerospatiale and MBB/Erno signed an agreement in December 1983 for the development of the SPACEBUS series (with weights ranging from 1 to 3 tons). Matra and BAE are investing FF 100 million each in a new family of light satellites called EUROSTAR. In parallel Matra is offering a small low-cost satellite, JANUS, for telecommunications, earth observations and scientific missions which could be launched as a piggy-back payload on ARIANE IV.

In Japan, consortia are much less firmly established and the manufacturing firms (Mitsubishi Electric, NEC and Toshiba, etc.) compete with each other in tendering procedures, each being associated with a US partner who provides technical assistance and/or components. Up to now a relatively low total volume of orders (see Annex I.18) and a reluctance to bear the costs of the investment required to develop individual techniques have discouraged firms from embarking on ambitious programmes. Thus, space represents 2 per cent of Melco (Mitsubishi Electric) sales and 0.5 per cent of Nissan Motors.

The recognition by the government of the need to improve technological capability for space activities and its determination to that end could lead to changes of attitude. There are already plans for a satellite, to be designed co-operatively by several Japanese companies, ETS-V. It is scheduled for delivery in 1987. Under the terms of the contract Melco, the prime contractor, will include NEC and Toshiba in the manufacturing of the satellite[61].

2. TRENDS IN INDUSTRY STRUCTURE

For the space industry, trends towards concentration or subsequent lessening will depend on changes in the relative importance of the factors of competition such as good growth prospects, specific government measures, satisfactory rates of return on investments, etc. and on the miscellaneous factors of co-operation such as capital-intensity of the activity, rising new technology acquisition costs and development of synergy. Although it is extremely difficult to assess these developments, it is possible to present a number of analytical points in respect to space activity sub-sectors where the situations are more homogeneous.

First, trends toward co-operation and concentration appear to be strengthening in satellite manufacturing. On the one hand, the various types of associations provide manufacturers with additional resources and possibilities of cost-sharing which help them win export markets. This arrangement is all the more satisfactory since demands from developing countries can be volatile and unreliable, especially within the present financially difficult climate, and the competition is fierce. On the other hand, procedures for allocating contracts to the multinational organisations tends to confirm acquired positions. Therefore, the number of producers in the world (20 or 30 at present) is not expected to increase in the medium term. In certain European countries the scale of concentration is already considerable. In France, for example, four producers have been responsible for two-thirds of the space projects implemented thus far.

For launchers it is unlikely that the present structure of supply for heavy lift capability will become less concentrated in the United States and Europe. Some other countries have developed their own launch vehicles, but it is still unclear whether these will be offered commercially (apart from the USSR's PROTON). On the smaller payload transport market, however, there may be room for new manufacturers. Some firms which are attracted by the profit to be gained on this market have already started the testing phase of their experimental rockets (e.g. SSI with the CONESTOGA, Starstruck with the DOLPHIN, Otrag).

For the ground station sub-sector, trends are less marked, or even divergent. Although the complexity of large stations may lead to the creation of associations or consortia, production is already highly concentrated in the United States and Japan. As an example of a multinational operation, 75 per cent of the INTELSAT stations were built by NEC and ITT. For the medium-size stations, frequencies generally determine the dividing line, with the United States dominating the 4/6 GHZ and the Europeans the 11/14 GHZ: each being master on their own ground to a great extent. Competition is therefore limited. Globally, these oligopolistic trends have been reinforced by the fact that producers of space systems are also builders of stations (e.g. Thomson through Telspace, the largest European supplier, SEP or NEC).

The opening of the market for small antennae could be the prelude to production activity by new firms. The strategy of the Japanese producers (60 per cent of the market) should contribute importantly to this development. The DBS segment should also provide good opportunities for many firms[62], perhaps even in the United States where the market is already dominated by a small number of firms: Scientific Atlantic claims 40 per cent of the market. In Canada local manufacturers have developed small, transportable, low-cost telephony earth stations and SARSAT local user terminals. In Europe the embryonic character of demand is likely to lead to gradual deconcentration and widening of the supply services.

These also appears to be a trend towards less concentration mainly in the United States for launch services and telecommunications, despite growing market entry cost (requiring start-up investment of at least $1 billion for common carriers). This trend is likely to affect particularly the space segment of specialised communications services. These are already subject to strong competition, and carriers can increase or redeploy their assets rapidly in order to seize profit opportunities. Because transponders, as well as their corresponding ground facilities, are widely available, no firm enjoys market power or has the possibility of fixing prices. According to an ATT study (1983), 11 companies could be supplying these services by the end of 1987 using 38 satellites compared to 8 companies using 26 satellites in 1984.

3. BARRIERS TO ENTRY

The development of concentration obviously brings up the question of the extent of barriers to market entry. These barriers, which are substantial for launchers, but less substantial for services, have a considerable impact on the satellite supplier's market.

The high threshold of initial fixed investment cost is the first entry barrier. This cost is so high for launchers that initially it can only be borne by governments. Entry costs represent the price of equipments and infrastructures needed to manufacture and test prototypes, and the cost of accumulating the knowledge necessary to master the basic, yet often very sophisticated, technologies. The learning curve in these sectors – several years for a satellite, at least a decade for launchers – naturally gives a decisive advantage to the first entrants in the field.

When markets are compartmentalised and producers operate in limited demand zones, the excessive costs due to the existence of short runs constitute an obstacle for new arrivals. In these circumstances, the supply of components and sub-systems necessary for the construction of a space product is particularly costly because the small number of units ordered increases the cost of manufacture.

At the same time, under present competitive conditions, new manufacturers seek to face competition by offering particularly advantageous financing arrangements in order to sell satellites and earth segment equipment[63]. Substantial price reductions are an alternative, but they are unlikely in a start-up period. While export credit arrangements to some extent limit excessive competition on financing terms, new formulas such as offset arrangements (which are becoming a commonplace customer requirement in the Third World) increase considerably the financial burden on the seller.

Finally, in essentially technological markets, technical excellence is a *sine qua non* condition of success. Thus no firm can establish itself as the architect of a project before having proven itself. Equipment manufacturers wishing to penetrate the systems and sub-systems market have also to meet severe technical constraints in terms of reliability of equipment (e.g. resistance to particular thermal conditions, the need to miniaturise sub-sets or to reduce overall weights). Stringent development and quality control for this kind of technical demand call for substantial expenditures. Even if the space quality label constitutes a valuable selling point for these products, many firms are unable to make such investments.

4. FIRMS AND TRADE

Given the limited information available, no comprehensive quantitative analysis of the economics of space products can be made. National statistics curently aggregate aeronautics and space activities and in the rare cases where they can be isolated, the classification adopted does not make it possible to distinguish between civil applications and military equipment.

Where trade is concerned, other limitations arise. Products within the space sector are often traded as subsystems rather than as final products. Therefore most of the information is not captured[64]. In addition, certain types of exchanges relate mainly to international technological agreements such as contributions in kind between countries. These practices are an important component of international co-operation and are valuable for both partners, but do not show up in trade statistics. Examples are the first Remote Manipulator System donated by the Canadian Government to the US Space Shuttle Programme and the first Spacelab placed by ESA at NASA's disposal.

Although trade should be interpreted with great caution given the information gaps, the following comments should shed some light on the topic:

a) The volume of trade in finished products between different producer countries in the OECD area and the rest of the world is relatively small, and as a percentage of production is probably, in most cases, below the corresponding levels for manufacturing industry as a whole. In France, exports of space equipment in 1981 amounted to approximately 15 per cent[65] of production, three-quarters of this being carried out in co-operation. In the United States the percentage of either propulsion systems or space vehicles exported remains small (see Annexes I.19, I.20 and I.21).

During the past few years the share of production exported by Japan has exceeded 20 per cent and is attributed mainly to the sale of ground station equipment. Japanese imports are at a similar level, but are concentrated in launchers, satellites and on-board equipment.

b) International trade in space systems depends largely on demand from developing countries and international organisations. Although these organisations base their choices on quality, price, delivery dates and arrangements for distribution of the products proposed, they must also take into account capital breakdown, i.e. between shareholders, usually public administrations such as national PTTs, or public companies. In general, international sub-contracting remains limited (about 18 per cent for the INTELSAT VI series, see Table 13). Leaving these international public contracts aside, the Third World alone offers numerous opportunities for world competition, especially for telecommunications satellites, as most national markets within the industrialised countries remain mainly captive, except for the earth segment.

c) With this market structure, firms concentrate their foreign transactions on the sale of sub-systems and components. At this level Japan's dependence on imported United States equipment varies between 90 per cent and almost zero for each of the 25 satellites launched by Japan so far[66]. In Europe the level of dependence is much lower (probably under 20 per cent in France). On the one hand, the national efforts made to increase autonomy relative to US technology may limit the prospects for trade growth in intermediate products at least between advanced countries. On the other hand, other factors such as the development of multinational co-operation and the broadening of the supplier base could create substantial trade expansion. To strengthen this expansion, international alignment of production standards would in any case have to come first, especially for sophisticated systems.

d) Traditionally, industrial firms try to cancel out the effects of barriers to trade by internationalising the location of their production facilities. It is noteworthy that in the space sector this has not been the case: the major producers maintain only commercial branch offices abroad. This special character of the space industry can be attributed to various factors which have already been mentioned such as the youth of the industry, heavy initial investment cost (in terms of facility cost and for

Table 13

Distribution of sub-contracting agreement[1] for the INTELSAT VI series

First group of five – first launch 1986

United Kingdom	BA	32.4
France	THOMSON	24.8
Italy	SELENIA	24.4
Germany	MBB	18.5
	AEG	5
Canada	SPAR	19.5
	COMDEV	1.5
Japan	NIPPON ELECTRIC	22.6
		($82 million)

1. The firm Hughes Aircraft is the prime contractor.
Source: Aviation Week and Space Technology, May 1982.

the team of skilled people required), still developing cost-effectiveness of non-US competitors, the power of government departments in the market, strategic character of certain activities, various regulations, special clauses in licensing agreements, etc.

e) In the past few years, trade has not only developed for subsystems and components, but for all segments of the finished products' market. This is especially true for earth stations (Graph 4). Japan, for example, has appreciably increased its trade balance since 1977 in this segment[67]. Also, many satellites have been produced in various countries for sale to other countries or international entities. Launchers are now considered as trade in services. Despite the limited supply side of this market, launch competition may become intense between Europe and the United States.

Graph 4 **INTELSAT STANDARD A EARTH STATIONS SALES BY SOURCE COUNTRY**

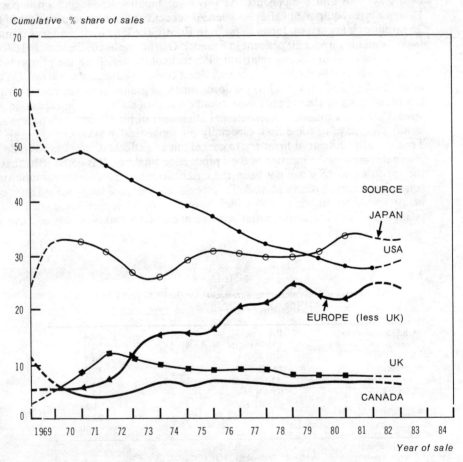

Cumulative % share of sales

Source : *Futures,* October, 1982.

48

Chapter VII

ISSUES RELATED TO GOVERNMENT POLICIES

Some governments recognise that broadly based national interests justify a substantial public commitment to space activities. This is especially true of R&D activities, but also applies to the development of launchers and certain satellite programmes. It has been pointed out that the scale and distribution of this government support somewhat determines the competitiveness of the different space industries. Canada, whose success on the export market is evident, bases its success on a consistent and comprehensive policy enabling the satisfactory development of a domestic market. The relationship is equally clear in the case of military expenditures and is likely to become even stronger than in the 1970s. For example the United States Defense Department budget in 1982 amounted to $460 million for telecommunications satellites alone. In contrast, NASA expenditures for these activities has been maintained at a low level in current years (e.g. about $25 millions in 1982). Clearly, the 40 or so military communications satellites[68] scheduled over the next few years constitute a very worthwhile safety net for US industry and even make it possible to extend production runs and hence reduce manufacturing costs.

Recently, two new elements have influenced the framing of space policies. On the one hand, changes in civilian budgetary policies in many countries for reduced overall public expenditure have increased the financial constraints and given rise to trade-offs, such as the dropping of certain scientific programmes in the United States and limitations on the diversification of projects, etc. On the other hand, in most countries, space activities as a whole still have a rapid rate of growth with favorable demand prospects. In the US, the supply of satellite communications systems remains buoyant (see Annex I.22) and private investment in commercial space projects is accelerating; the latter rose from $10 millions in 1980 to more than $175 millions in 1983[69]. In Europe, thanks to the commercialisation of the ARIANE booster and to the proliferation of telecommunications projects, the industry is experiencing a boom (see Annex I.23). European firms estimate that by the mid-1990s the market for their products will have doubled compared to now[70].

However, for certain segments of the space market, trends are still shrouded by uncertainties. For DBS, progress is slower than anticipated. In the US, some companies (CBS, Western Union) withdrew from the market and others curtailed their investment projects. For space telecommunications, a strong challenge may be expected in the long term from fiber optics cables[71]. Even if the ATT TAT-8 undersea fiber cable planned for the end of the decade does not significantly affect the satellite market[72], the next generation of optic fibers may offer considerably greater capacity at lower cost and therefore may threaten the competitiveness of satellite systems (at least for heavy routes). In the area of materials processing in space (MPS), SPACELAB and retrievable carriers such as EURECA from ESA or LEASECRAFT from Fairchild will provide scientists with increased opportunities for further experiments. However, at present, MPS commercial viability is not established. It will depend

49

on how many improvements can be obtained by manufacturing in space versus alternative technology development on earth and on the cost of space flights and materials transportation.

Governments are accordingly seeking to:

i) increase the proportion of internationally co-operative programmes in order to reduce financial burdens. Now that the Shuttle is operational, as a next logical step in the US space programme, NASA plans to build a permanently manned orbiting space station (for a possible launch in 1991). The United States Government has invited Canada, Japan and ESA Member States to participate in the development of this project and technical discussions are being held with NASA. The cost of the programme is in the region of $7 to 9 billion. In Europe, ESA countries recently agreed to start the first phase of a long term programme based on a new orbital infrastructure – the COLUMBUS space module – and a launcher with heavy lift off capabilities – ARIANE V[73];

ii) particularly encourage private investment in space using incentives and therefore increase industrial opportunities. Efforts to secure equal tax treatment to space endeavours versus terrestrial investment and to promote an appropriate regulatory environment are receiving special attention. In the United States an office of commercial space transportation has been created within the Department of Transportation to co-ordinate unmanned launch vehicle operations. In addition, an office of commercial programmes was established in NASA and a working group, chaired by the Department of Commerce, under the Cabinet Council on Commerce and Trade was formed in response to the President's 1984 "State of the Union" address, calling for increased emphasis on commercialisation.

Discussion about the commercialisation of space activities is becoming widespread and is centred on a number of precise themes. Most of the following issues have a direct or an indirect bearing on the trade of space goods and services.

1. TRANSFER OF ACTIVITY TO THE PRIVATE SECTOR AND/OR TO USER CONTROL

Satellites

The United States government has decided to transfer the responsibility of the LANDSAT system for terrestrial observation[74] – remote sensing by satellite – presently managed by NOAA, to the private sector. LANDSAT privatisation has been recently approved by the United States Congress. The enabling bill provides for phases of commercialisation starting with the selection of a contractor to market all new and previously archived data from the present system. Operating contracts will also be permitted. The next phase, lasting six years, is considered a transition period during which the contractor will be paid a government subsidy to maintain system and data collection continuity.

A first request for proposals was issued by the Source Evaluation Board of the United States Department of Commerce and the administration has proceeded with calls for bids. In the RFP NOAA requested bidders to abide by the principles of non-discriminatory dissemination of information. Successful bidders were narrowed to two firms: EOSAT, a joint venture of Hughes and RCA, and Eastman Kodak heading a team that includes Fairchild and TRW as subcontractors. Ultimately, EOSAT was designated as the commercial operator of the LANDSAT remote sensing spacecraft.

50

This sale raises some issues. For example, it has been said that the proposals by the private sector will be subjected to conditions regarding guaranteed contracts (buying of data by public administrations). In addition, one of the two satellites in operation (LANDSAT D) still has technical difficulties and thus increases the risks that the selected bidder encounters when taking over the program. Simultaneously, users of remote sensing satellites would appear to fear that privatisation[75] might cause sharp rises in prices. Lastly, the question arises regarding the monopoly a private company could enjoy in the market for remote-sensing data, especially if the company owning the data collection and distribution system was also allowed to offer value added services.

In Europe the programmes planned have not yet culminated in the actual launching of satellites (except for telecommunications) but major progress has been made in transferring the responsibility for managing existing satellite systems to user organisations, i.e. EUTELSAT and EUMETSAT, and for specifying and procuring their future satellite requirements. In France, too, a semi-public management company, Spot-Image, has been created which is responsible for distributing the data supplied by the SPOT system. This company is hoping for a large demand from foreign markets equalling nearly 80 per cent of total demand for the system. Spot-Image is presently expanding the number of agreements signed with customers, who will receive the data either through an agent of Spot-Image or directly through their own earth stations.

Launchers

Privatisation of ELVs has also been part of the US space policy, since the President announced in May 1983 that the government will encourage and facilitate this aspect of commercialisation; hence, private companies will be allowed to assume production and launching of ATLAS/CENTAUR and DELTA ELVs at the close of the government's programmes in mid-1986 and 1987. As an incentive, NASA and Defense Department launch facilities will be made available to the contractor at nominal costs, not at the cost of paying off the whole of the investment incurred. In Europe, support has been given to the Arianespace Company. ESA grants the company use of the Kourou launching site for a fee, based on nominal costs and exemptions from research costs for ARIANE 3 and 4, which are being covered by the organisation.

2. COMPETITION POLICIES AND PUBLIC PROCUREMENT

In the United States, deregulation (see below section 7) has become the preferred solution to promote the development of the satellite communications market. For satellite equipment and components, access to the market is completely open and consequently, the competition policy of the Executive branch is focused mainly on the launch market. While ELV commercialisation is encouraged, the STS is still considered the primary launch vehicle for the United States government and will continue to be available for domestic and foreign users. Full account is being taken of the fact that new ELV competition may affect the financial management of NASA's Space Shuttle.

To improve the STS programme efficiency, and to facilitate the involvement of the private sector in the operation of the Shuttle, launch processing at the Kennedy Space Center is now being managed by a private company under contract to NASA. This contract was won

51

by a consortium headed by Lockheed Services Inc. In addition, NASA has issued a tender for Shuttle marketing support[76]. Lastly a new company, C2 Spacelines, signed a memorandum of understanding with NASA for leasing the complete Shuttle cargo bay for a limited number of launch slots in 1987 and possibly thereafter.

Open bidding on NASA programmes is required by law. NASA considers offers from any source based on price, quality and the capability of the contractor providing the offer. At present the lead gained by US enterprises, both in time and experience, means that it is difficult for foreign firms or consortia to underbid them. Foreign industry has not played a major role in direct contractual relations with NASA. However, NASA has participated in international programmes and this has provided opportunities for firms from other countries to benefit from its activities.

In most countries, other than the United States and Canada, the demand for satellites comes mainly from government entities. It is primarily confined to a small number of producers, usually domestic, who serve as prime contractors for systems and sub-systems manufacturing. These producers are often approached because of their strong competitive position or because no other competitor exists; yet in a few other cases it is because policies of national preference aim primarily at the creation of industrial and technological capabilities. The grouping together of national producers, for example, has been encouraged in the United Kingdom (United Satellites for the construction of UNISAT). In Japan, MITI and STA (Science and Technology Agency) recently proposed to do research on part of the mission equipment for the Japanese ERS I (Earth Resources Satellite) satellite through joint efforts by several firms[77].

The situation is different for multinational organisations, because governments tend to establish links between their own contributions and the spin-offs for their national industries. Contracts for the construction, integration and operation of satellite sub-sets are therefore allocated among international participants whenever possible. This is the case regarding tenders for the various INTELSAT series. So far, however, the prime contractors have always been the United States and United States' industry – in particular Hughes Aircraft for INTELSAT I, II, IV and VI – continues to dominate the market. It may nevertheless be asked whether these procedures do not lead to certain constraints in the choice of manufacturers and therefore to a compartmentalisation of markets and trade. This could also be true of the European Space Agency, where the "overall return coefficient" (the principle of fair returns on national contributions) has been more stringently applied in recent years. It has, moreover, been said that this practice tends to penalise industry in the more competitive countries.

The need to move towards an open competitive procurement system has been recognised. Efforts to fulfil this goal will benefit both suppliers and users of space products. It will also permit an expansion of trade opportunities and allow countries to fully specialise in equipment and services according to their comparative advantages.

3. LAUNCH SUBSIDIES AND THEIR IMPACT

Estimations of launch costs and prices can vary according to demand forecasts and the time deemed necessary to recover the investment (see Chapter IV section 3). Nevertheless, whatever the calculation, it is generally admitted that launching services are subsidised either directly or indirectly. The difference between launch costs and prices may however be small relative to total costs and to revenues derived from telecommunications services. In the future, cost per flight may decrease (in volume terms) depending on the progress made in increasing launch effectiveness, or the growth of satellite demand which in turn increases launch demand. At the same time, prices will increase relatively with the gradual phasing-out of subsidy policies and will make up the difference with costs. The question during the transition period is whether these conditions will not penalise private services, using conventional rockets, or create distortions in the competition between satellite and cable transmission systems. Further investments in space communications may conceivably be considered as creating overcapacity.

4. THE STANDARDISATION ISSUE

Lack of compatibility and interchangeability of products is a barrier to trade which should not be underestimated. On foreign, intermediate or end-user markets, failure to comply with prevalent technical standards hampers competitiveness. Thus it is generally admitted that US satellite builders are constrained in many foreign markets, mainly in the Third World, because of differences between ITU and US domestic specifications.

Until now, world standard setting has been developed by international organisations, and mainly diffused and elaborated upon via the ITU's non-binding recommendations. INTELSAT also plays a leading role in promoting technical standards, especially for earth stations, due to its existing network and extended participation in international telecommunications. For the space segment, it contributes to the harmonisation of hardware through the design of new spacecraft and the involvement of numerous countries' firms, which have INTELSAT procurement contracts.

The need for harmonisation has also been recognised at regional levels. In this regard, CEPT recently formulated guidelines for the establishment of common certification procedures. A group will be created to set up European standards with a view towards making them compulsory. In addition, the validity of technical tests performed in one country will be extended to others.

Co-operation agreements between market leaders could also result in the setting up of European standards. For example, the 12 leading European computer and communication manufacturers recently decided to co-operate in the application of common international standards. Furthermore, the two largest producers of TV sets, Philips and Thomson-CSF, have agreed to set up a common standard (D2-MAC package) for DBS which might eventually be adopted throughout Europe.

While the development of standards and certification systems on the largest possible scale can significantly help to facilitate trade opportunities, they should not be applied in a manner which would constitute a means of arbitrary discrimination or a disguised restriction on international trade.

5. SPECIFIC IMPORT REQUIREMENTS

Due to the nature of the space industry, many OECD and non-Member countries do not plan to develop a completely indigenous system supply capacity, but to cover the high costs involved in purchasing, they have asked for a variety of industrial and financing arrangements from bidders such as offsets, parallel financing and other forms of assistance. In particular, they often require local production of a given fraction of industrial inputs. Governments are generally involved in the negotiations. This can be advantageous to those countries where public authorities have traditionally played an active role in industrial affairs.

Tenders for large markets in non-producer countries or regions, for which consortia from supplier countries compete, will hence generally include financial and export credits, local sourcing of industrial components and transfers of technology. Competition can be waged principally at this level. Since governments sometimes have important stakes in the industry, they may seek to play a role in the negotiations following the publication of tenders.

6. AID FOR R&D

The stakes in the space game will be high notably with the possibilities offered by industrialisation of activities in orbit such as the manufacture of pharmaceuticals, biological products and semi-conductor crystals, e.g. Gallium Arsenide, and experiments in metallurgy. However, the industrial sector is devoting few resources to space R&D because of the need for more basic scientific knowledge, the risks involved, the long time needed to recoup investment outlays and the high costs to be borne. Consequently, governmental space agencies continue to carry out much of the research themselves or to subcontract it to industry. In certain countries, activities have even increased considerably in recent years (for example, CNES research appropriations in France increased by 56 per cent in 1983) whereas in other countries support for civilian R&D may remain at current levels or decrease in volume.

Resources in the United States have been made available to increase funding for research in related sectors with space applications, such as robotics and semi-conductors. Supplementary funds were provided in 1984 for robotics and automatic systems by the research body of the Defense Department (DARPA) or the NSF. In Canada funds for robotics were used to develop the RMS (Remote Manipulator System) for the Shuttle.

Moreover, space agencies in the United States, Europe and Japan continue to fund scientific programmes, such as astronomic observatories and space science laboratories. Even though these programmes represent diminishing proportions of budgets, they are still by no means negligible (15 per cent for NASA, 14 per cent for ESA and 10 per cent for the National Research Council of Canada). The induced effect on R&D is important as manufacturers can innovate and take technological risks (in contrast with commercial contracts where competition allows no mistakes).

Governments do, however, want to involve increasingly the private sector in research activities on a shared cost basis. One example is the initiative taken by NASA to associate manufacturers in the Ka band (30/20 GHz) satellite communications demonstration project, ACTS. In the space telecommunications market, most profits have been made by carriers (those who lease transponders to users) rather than by constructors, so that the latter can find certain advantages in accepting NASA's formula.

The Agency has also established a procedure for Joint Endeavour Agreements (JEA), under which US signators commit funds for the testing of new techniques for materials processing. Among other resources, NASA may provide flights aboard the shuttle at no or reduced cost and allow use of its ground-based facilities to support private sector JEA proposals, where significant capital is at risk and the proposed activities contribute to the goals of the United States space programme. Private contracting companies exercise some proprietary control over the information about the results. However, these companies may be obliged to share data and/or results with NASA as part of the *quid pro quo* of the agreement. Examples of JEAs signed thus far are the agreement on electrophoresis[78] (with McDonnell Douglas/ Johnson and Johnson), on organic crystals (3 M corp) and on Gallium Arsenide crystals (Microgravity Research Associates). NASA broadened the scope of JEA activities by concluding a JEA with Fairchild Industries to develop a retrievable carrier, LEASE-CRAFT[79]. Several other JEAs are being considered.

In Europe, no such procedures have been implemented and the private sector is less involved in R&D funding. For example, the EURECA platform will be entirely financed by ESA. However, in the case of equipment with short or medium term commercial potential, European industrialists may invest their own funds. The EUROSTAR platform is thus developed over 50 per cent by British Aerospace and Matra.

7. DEREGULATION

The determination to reduce the risks of commercialisation of space services and to relax restrictions on its development has led certain countries to abolish many regulations which are considered "inhibiting". The objective is to stimulate competition and enable new carriers to enter the market. Since regulations vary greatly, only a few examples can be provided herein.

First, in 1981 the FCC and the United States Congress deregulated the international carrier market. International carriers were thus authorised for the first time to offer their services to the US market. In addition, the distinctions made between the transmission of computer data, the telephone and telex systems were abolished. Nevertheless obstacles still prevent unification of the entire market, because the postal and telephone services of the destination countries sometimes refuse to sign agreements with newcomers on the US market.

Secondly, in June 1982 the FCC authorised direct television broadcasting and the launching of corresponding DBS satellites. Programmes will be broadcast directly to homes equipped with suitable antennae. The first links were set up at the end of 1983. In Europe, France and Germany launched their DBS project TDF1/TV Sat in 1979 but these satellites will not be in operation, together with ESA's OLYMPUS, before 1986/87 (later for the British UNISAT). For the time being television in Europe is performed by the ECS satellites (and soon by TELECOM 1) and relayed through intermediate earth stations and a cable network. These networks are not generally covered by the deregulation and continue to be under the control of the postal and telephone systems. When the various DBS services start, however, controlling conditions will change and competition will increase.

Deregulation should allow rapid growth in demand for small antennae and ignite sharp competition among producers. Substantial price reductions are expected.

Thirdly, the principle of free collection and non-discriminatory dissemination of pre-processed data supplied by remote sensing satellite operators continues to be adhered to. The application of this principle will not only secure the interests of the customers, but should also provide a regulatory environment which is conducive to international competition. Despite the present high commercial risk of the remote sensing data market, the transfer of the LANDSAT system to the EOSAT company may lead to the creation of Earth data service markets specifically tailored to customer needs. In its wake, new imaging ventures as well as other remote sensing operators should enter the business. Trends will also depend on progress made in the supply of low cost data processing terminals. Already, the Spot-Image company created in 1982 is setting up an international network to sell images which will be supplied by Frances SPOT satellite. Prices will be calculated in order to recoup management and recurring costs.

Fourthly, within the liberalisation of the United States market for telecommunications services, proposals to relax rules governing the operation of the public firm Comsat, which looks after United States interests in INTELSAT, are being studied. One such proposal aims to let circuit users other than common carriers deal directly with Comsat and authorises the latter to provide services to the public on the basis of cost-related prices as opposed to current fixed rates. The objective is to establish normal competition between Comsat and other carriers. In addition, the current ownership system for United States INTELSAT ground segment has recently been revised to enable users to purchase earth stations under certain conditions.

Outside the United States, steps have been taken in some countries to reduce the burden of administrative and regulatory constraints. In Japan the near monopoly of telecommunications by NTT (Nippon Telegraph and Telephone Public Corporation) is being reformed in order to promote private capital investment in the telecommunications industry and thus boost competition. Private enterprises will be given the opportunity to utilise satellite transponders for themselves. In the United Kingdom splitting up this sector was considered impracticable[80]. Nevertheless, part of British Telecom's capital (up to 51 per cent of shares) was recently transferred to the private sector. The policy for selecting space equipment suppliers may be affected.

As deregulation increases, acute problems are raised. It is difficult to draw a line between public or semi-public and private activities. Furthermore, many international treaties or agreements signed by governments may reduce the latter's room for manoeuvre. The recent conflict between INTELSAT and the leaders of the ORION and INTERNATIONAL SAT projects[81] is a good example. In principle, the non-INTELSAT satellites for international linkages are governed by INTELSAT's rules. They state that companies providing alternative services must supply proof of technical compatibility between the two services, and provide evidence of the absence of economic and financial detriment to the Organisation. The discussions about the implications of the two projects included the volume of diverted traffic that might ensue (less than 4 per cent according to one of the parties), the possible effects of expansion on the market through the introduction of new services and lower prices, and abolition of the less profitable international routes in the context of progressive deregulation of links. Since early 1984, other new projects have emerged: a number of companies have applied to the FCC for authorization to offer an alternative to INTELSAT services. At the end of 1984 the Executive Branch agreed to allow competition in international communication satellite systems under certain conditions (such as a ban on these systems carrying switched services and the necessity for competing firms to seek international approval through INTELSAT coordination).

Under the GDL project, Luxembourg is planning to launch a new TV satellite system, CORONET. By using two US built satellites this system will channel TV programmes to cable networks in Europe and to houses equipped with TVRO terminals. On completion this project will challenge EUTELSAT's quasi-monopoly for intra-regional communications.

Chapter VIII

CONCLUDING REMARKS

A preliminary result of the study has been to emphasize the need for a more accurate definition of the space industry and for the building of a sound statistical base, particularly in the area of trade[82]. This is clearly a prerequisite for further analysis.

The development of the space products industry and the expansion of international trade in space products cannot be dissociated from the original features which have marked space activities. Without extensive governmental involvement, the development of space activities and resulting industrial products would never have occured. Strategic considerations were of prime importance at the birth of the industry and still remain a decisive driving force.

In considering the future development of the industry and the expansion of trade, it is best to recognise that governments will continue to play an important role in funding the development and production of space products, as well as regulating space activities (most of which have an international dimension and international implications). Any common perspective of the industry, the expansion of trade and the role of governments must also be based on the understanding that countries undertook space activities at various periods and are still at different stages of development.

The conditions for the development of an open multipolar trading system have only gradually emerged. A truly international demand for civilian space products, in particular civilian telecommunications satellites, launchers and ground stations, has begun to develop over the past twelve to fifteen years. There is now an expanding international market for these products.

Until about five years ago the United States was almost the exclusive exporter of space items. Recently, the supplier base has grown and become international. The supply of satellites is still largely dominated by United States industry, but firms, organisations and government agencies in other countries are increasingly entering in the field. Launch vehicles are now available in Europe and Japan as well as the United States. Ground systems are perhaps the most internationally competitive of all the space related products, with suppliers in Canada, Japan, Europe and the United States. The sourcing of components, in particular for the electronics capital goods industry, has taken place on an international basis for many years. As the space products industry evolves from the development phase to broader commercial utilisation, the volume of trade will increase. The strengthening of an open multipolar system of trade is still, however, an objective to be attained.

As noted in this study, government roles have gone far beyond sponsoring R&D and advanced research, since they have been involved in space activities as a regulator of the market, as a customer and as a supplier of equipment and services. Their involvement has depended on predefined economic structure. Given the increasing exposure of markets to international trade and the broadening of the supplier base (e.g. telecommunications services, earth stations manufacturing, satellite components, etc.), the trade implications of domestic policies will become more important. With telecommunications deregulation extending in some countries, competition for domestic and global services is growing and the question of the necessity of retaining international monopolies may arise. The expansion of satellite communications systems also represents a strong pressure towards transforming the traditional organisation and operation of telecommunications systems in a number of countries.

In the space products industry, the move towards more competition and greater multipolarity will remain closely associated with, and complemented by widespread close scientific, technological and industrial co-operation between countries and firms. In Europe, the establishment of ESA has helped smaller countries and firms to partially overcome the disadvantages of size and thus participate in the space adventure. Ties between the United States and the Canadian space industries have always been close and Japan has considerably benefited from US transfers of space technology. The Shuttle flight of the Spacelab and the present discussions on the conditions for integrating the COLUMBUS module within the US space station examplify the dynamism of American-European cooperation. The offer made by the US to other countries to participate in the construction of this space station will create new opportunities for the further development of such co-operation. Finally, the decision taken at regional level by the ESA countries to undertake the first phase of a major long-term programme based on the development of ARIANE V and COLUMBUS, and the interest shown in the Space Aircraft HERMES project, presage a strengthening of the cooperation within the European industry.

NOTES AND REFERENCES

1. The first flight of a liquid motor rocket took place in Massachussetts (US) in 1926 as a result of Goddard's work.

2. Because of the decisive contribution of NASA, the introduction of these satellites on the market was probably advanced by several years. It should be recalled that considerable reservations were then expressed by industry and, in particular, by ATT (which had already launched the international television satellite TELSTAR) about the geostationary technique, in spite of the extra costs of using non-synchronous satellites (which required a chain of satellites to maintain contact with earth and large investments in earth stations and tracking systems).

3. It took over the activities of the National Space Development Centre set up in 1964.

4. Several studies demonstrate the importance of spin-offs to manufacturing industry, such as NASA's report, "Economic impact of stimulated technological activity", October 1971, which predicted a return of 7 to 1 up to 1987 on all space expenditure between 1959 and 1969. In Europe, return on ESA spending is evaluated on average at 3 to 1 according to a Strasbourg University Report (June 1980). It must however be emphasized that most of these studies ignored opportunity cost considerations.

5. The most remarkable fields of application concern mastery of information problems (with the development of new technologies for telecommunications, e.g. travelling wave tubes, information storage and optical techniques), energy questions (solar generator, heat control), transport systems and management and organisation methods.

6. Although ELDO never succeeded in making the Europa rocket operational, European co-operation was tried and tested in the construction of satellites by ESRO.

7. The organisation has been making profits between 1973 and a recent period.

8. A Soviet military satellite has a much shorter lifespan than its American counterpart and must therefore be replaced more often. In addition, differences in technologies and missions limit the scope for quantitative comparisons.

9. These satellites use variable orbits, geostationary at 36 000 km especially for telecommunications (the satellite is fixed with respect to Earth), lower polar or heliosynchronous orbits for observation and weather forecasting.

10. For the US Space Shuttle, the orbiter must be regarded as one stage.

11. This can be identical even for different payloads.

12. The electronic device performing the amplification of the received signal and the translation of frequency (necessary to avoid interferences) is called a transponder. It is a capacity element which handles simultaneously numerous circuits for the telephones or other types of communications. Two transponders are necessary to create a bilateral link.

13. Formed by the European Conference of Postal and Telecommunications Administrations.

14. See Baker (November 1982) or EUROSAT (December 1982).

15. i.e. about 80 per cent of telecommunications traffic.

16. See M. Savage, C. Catoe and P. Caughran, "Manned space station relevance to commercial telecommunication satellites: A prospectus to year 2000", AIAA/NASA Symposium, July 1983, Arlington, Virginia, US.

17. The relative growth in the demand for satellite communications, revised downwards recently, is nevertheless expected to amount approximately to 10-15 per cent in real terms annually.

18. See "La lettre de conjoncture", Banque Nationale de Paris, November 1983.

19. For an idea of transmission rate differences, an ordinary telephone transmission requires about 2 400 bauds, bits per second, and a teleconference 1 million, making satellite far more attractive than cable.

20. See Marc Giget, *L'espace commercial : enjeu des années 80,* Aeromag, July/1982.

21. The INTELSAT Organisation owned and operated at the end of 1983, 17 satellites, from which it earned that year $366 million. For the last ten years or so, INTELSAT has averaged approximately 16 per cent return on capital invested.

22. INTELSAT is currently leasing to INMARSAT four MCS (mobile communication service) sub-systems. It is also undertaking studies to explore how future INTELSAT spacecrafts might provide aeronautical mobile communication services. These studies could lead to a satellite system concept, whereby INTELSAT might offer land mobile services. See INTELSAT: The Global Telecommunications Network, Pacific telecommunications conference, January 1983, J. Pelton, M. Penos, A. Sinha.

23. IBS (Intelsat Business Service) is fully digital and carries all types of telecommunications services including video teleconferencing, high and low speed facsimile, high and low speed data, jacket switching, voice telephony, electronic mail, telex, etc.

24. For example, SBS was authorized in 1984 by the Federal Communication Commission (FCC) to operate as an international carrier with the United Kingdom (for data transmission, teleconferencing, facsimile, etc.).

25. Two US businessmen have established the ORION project with financial support from industry. Their plan is for NASA to put two satellites into orbit during 1986 and 1987 for intercontinental private lines services between North America and Western Europe. Reservations have already been made with NASA, but ORION has not yet received a formal approval. It has been argued that this project may threaten INTELSAT's market over the North Atlantic route and could affect its financial status. The organisation derives much of its revenue from this route.

26. Countries involved include Canada, France, the United States and the USSR. Norway and other countries have joined as investigators.

27. In addition to advantages resulting from smaller size supply, purchasing of transponders by users (1 transponder is equivalent to 2 000 circuits for a GALAXY for example) allows stabilisation of fees and eliminates the risk of service interruption in favour of higher priority customers.

28. Transponder rentals in the US generated revenues estimated at $400 millions in 1983 by International Resource Development Inc. These revenues could rise to about $1 billion in 1988.

29. SBS and Geosource formed a joint venture (50/50) GEO/SAT communication. Its aim is to provide telecommunications services to the oil industry, especially for links between offshore platforms, data processing centres and office headquarters. SBS and/or a WESTAR 5 will provide these links.

30. In this context, note the capacity of SBS 1 450 million bauds, or the equivalent of 200 000 telephone circuits.

31. More than 200 watt, i.e. approximately 10 times the power of a conventional satellite.

32. Apart from Satellite Television Corp. (STC), there are numerous contenders for this market, e.g., DBS Corp., US Satellite Broadcasting (USSB), Dominion Video Sat. Inc., Hughes Communication Galaxy, National Christian Network, Advanced communications Corp., Satellite Syndicated System and probably in the near future cable programmers such as Home Box Office (HBO) and Turner Broadcasting.

33. A first experiment has been performed in Indianapolis by USCI. It started in November 1983 using one Canadian ANIK. The broadcast service ended recently. Installation charge were $300 and subscription including equipment rental amounted to $40 against $20 to $30 for cable TV. In addition STC should start a temporary quasi DBS service over the North-East part of the US using some SBS4 transponders. This service will normally end when its DBS satellite becomes operational.

34. Launching of the first high definition satellite UNISAT, scheduled for 1987, will certainly be delayed for one to two years.

35. National systems have also extended to developing countries, e.g. Indonesia's Palapa and India's INSAT satellites. Projects to be implemented between 1984 and 1986 are for the Arab League's ARABSAT, Brazil's BRASILSAT and Mexico's MORELOS. As regards the OECD area, the Australian AUSSAT project culminated in a 1985 launch.

36. At the June 1982 Versailles Summit, this was the subject of a United States proposal for technological co-operation.

37. The resolution power or sampling rate achieves 30 metres with the new thematic mapper system of LANDSAT IV. It is expected to be 10 metres for SPOT.

38. As far as Canada is concerned earth station manufacturers have focused mainly on producing INTELSAT standard stations, low-cost transportable telephony terminals, LANDSAT remote sensing terminals and SARSAT local user terminals, and TVROs. The latter can probably now be considered as consumer electronics products, with numerous countries attempting to develop production capabilities.

39. Because of their strong competitive position, Japanese manufacturers have penetrated deeply into the American market (in particular accounting for half of the SBS network).

40. Estimate by Euroconsult.

41. Already about 1 million American homes are equiped with circular dishes suited for the reception of low power DBS. The number of these private dishes is increasing rapidly with 25 000 to 50 000 more each month. However some changes will certainly occur as companies such as HBO have decided to scramble their service.

42. See Mackintosh consultants: "Satellite broadcasting".

43. Early Bird (1965) had a maximum capacity of 240 telephone circuits; INTELSAT VI (1986) will provide some 36 000 two-way telephone circuits or the equivalent of 120 colour TV channels.

44. Because of the growing number of two-way circuits available, the price of a half-circuit has constantly declined from $32 000 in 1965 to $4 680 in 1980.

45. For a conventional telecommunications satellite, extra annual earnings of the order of $0.8-1.5 million can be expected for each transponder kept in operation.

46. Considering the space market as a whole, the addition of the NASA budget and that of the Defense Department for space gives an estimate, for the US market, of nearly $15 billion in 1983, as against $1 billion for the European market ($700 million for the ESA budget, plus $300 million for country programmes excluding national contributions to ESA) and $0.5 billion for the Japanese NASDA budget. The ratios are therefore of about 1 to 15 and 1 to 30. By comparison, the Canadian government's space budget for 1983 is about $109 million.

47. In 1974, a 10-metre antenna with uncooled parametric amplifier front and operating at 6/4 GHz cost $125 000. In 1979 the same antenna with an uncooled field effect transistor front end was sold for $60 000. See "Telecommunications, Pressures and Policies for Change", OECD, 1983.

48. The US TDRS (Tracking Data Relay Satellite) has shown the way here. This spacecraft is a body stabilized vehicle whose multiple antenna provide two-way communication and data services for the space shuttles and for low altitude spacecraft. TDRS-A, the first of a series of three, was launched in April 1983 aboard the shuttle. Although some technical problems linked with the IUS module were experienced during its placement into orbit, TDRS is now working satisfactorily, as illustrated by the first Spacelab mission. This space system allows the transfer of vast amounts of data from low orbit to the ground. It can therefore substitute for existing ground-based network of tracking stations used by the United States to support its near earth space activities.

49. The range of insurance depends also on satellite and apogee and perigee (if any) motor performance. Since 1984, satellite insurance premia have considerably increased.

50. About 30 per cent of the Shuttle's payload has been reserved by the Defense Department.

51. The cost on completion in 1985 of the entire Shuttle programme amounted in 1982 to $16 billion. This covers the combination of development costs and construction expenditures for four orbiters ($1 billion each). For the ARIANE Rocket, FF 8 billion has been invested up to now in the basic development programme and follow-up to ARIANE 3.

52. Fee for a PROTON launch was listed by the Soviets as approximately $24 million in current year dollars.

53. For a geostationary orbit the Space Shuttle, unlike a conventional rocket, does not give its satellite sufficient momentum when liberating it into transfer orbit and it therefore requires a module with additional motors (perigee motor). For a payload of under 2 tonnes (in orbit net value) the McDonnell-Douglas PAM (payload assist module) is generally used. For a very heavy load (up to 6 T), NASA has ordered a small number of the CENTAUR stage constructed by General Dynamics and Pratt and Whitney. Utilisation of this stage necessitates alteration of the ground infrastructure (for an estimated cost of over $100 million). For an intermediate load, USAF and Boeing have constructed a propulsion stage called IUS (Inertial Upper Stage). As mentioned above, its operational reliability has not yet been proven. Malfunctions during mission 6 compelled NASA to review its schedule and to cancel one launch. These configuration questions have an immediate impact on demand for launch services and also affect the pattern of supply. Alternative solutions are being proposed. Astrotech/Mc Donnell Douglas is developing a DTS stage based on DELTA's second stage and RCA its SCOTS (Shuttle Compatible Orbital Transfer Subsystem) module. To meet similar requirements OSC/Martin Marietta is preparing the TOS/AMS Transfer Orbit Stage combined with an apogee motor to kick IUS class payloads into orbit.

54. Arianespace has an order book of $0.8 billion (of which 40 to 50 per cent are for exports outside EEC) to launch more than 35 satellites and hopes to take one-third of the world market. Between 1982 and 1986 the Shuttle market is expected to be worth $2.5 billion, and $4 billion from 1986 to 1991.

55. In France, the United Kingdom and the United States the aircraft industry as a whole received in average, around 50 per cent of the total public R&D assistance allocated to the manufacturing sector during the 1970s.

56. US military orders, for example, represent several billions dollars a year (more than 5 in 1983). In Europe the United Kingdom and France spend important sums on military programmes (inter-continental missiles), whose technology has many points in common with space technology. The United Kingdom operates a programme of military communications satellite programme called SKYNET, while France has adjourned its SAMRO observation satellite project. However for Western Europe as a whole, combined military spending for space does not exceed 2 per cent of the total space budget.

57. Analysis will be confined to suppliers of the space segment (attitudes are much more uniform in a market that is as exposed to international competition as ground stations).

58. Hughes sold over 30 HS 376 platforms.

59. Revenues of US satellite common carriers should increase from an estimated $100 million in 1980 to $1.2 billion in 1988. See Electronic Industry Association, Electronic Market Data Book, Washington, 1984.

60. Spar's total revenues including space product sales in 1981 were about $98 million.

61. This association was attractive on account of the possibilities for standardization of the platform and the variety of payload functions. The satellite should make it possible to link adequately equipped aircrafts and ships with ground radio stations.

62. Agreements for the manufacturing of related antennas have already been signed between companies, such as the NEC/ALCOA agreement.

63. Certain analysts consider that the participation by banks in the capital of the Arianespace company constitutes a significant advantage for the European rocket.

64. For example, US import statistics do not reflect import of components for the INTELSAT contracts.

65. *Source:* GIFAS.

66. Japanese firms limited investment in space activities and the absence of any military programmes are arguments currently put forward to explain the country's dependence in this area of technology.

67. See data from the Society of Japanese Aeronautic Companies (SJAC)

68. Ten satellites for the DSCS III system, three for FLEETSATCOM, 28 NAVSTAR communication and navigation satellites (GPS) and two heavy LEASAT satellites, at least.

69. According to the Center for Space policy.

70. See ESA bulletin February 1984: "The Outlook for World Space Expenditure", G. Dondi and M. Toussaint and Eurospace: "Toward a long term european space programme", 1985.

71. By 1987, US consumption of fiber optic cables could reach $700 million compared to $200 million in 1983 according to Electronics Magazine.

72. While the capacity of the INTELSAT VI is 1.6 (operational configuration) to 3.6 (maximum capacity configuration) times greater than TAT 8, the annualized cost of the satellite system (including earth stations) is estimated 12 % higher than the TAT 8 cable system. See "Satellites versus fiber optics cables" by Cummins, Lemus, Reyno and Crispin. Paper prepared for the Pacific Telecommunication Council Conference 1984.

73. The COLUMBUS programme (including the manned module, the platform and the preparatory missions) and the space transport programme (including the development of the ARIANE V version and the HM 60 motor) will cost each approximately FF 18 billions.

74. As a result of legislation, the management of the weather forecasting satellites will remain in the public sector.

75. While the entry into space business is generally referred to as commercialisation, a distinction between privatisation and stricto sensu commercialisation should be made. Companies taking control of government programmes such as LANDSAT or ELVs are involved in a privatisation process. It should be noted that, in this case, most of the risk is borne by the public sector. Commercialisation ventures refer to firms committed to new developments which have higher risks such as materials processing or advanced telecommunications.

76. One company is also prepared to market the Shuttle through purchasing of one orbiter, but NASA and the Congress have not yet announced any decision regarding this sale.

77. The method and budget are subject to change by the Authority.

78. This is a technique for the separation of organic substances in solution under the effect of an electric field. Its efficiency in space is estimated to be over six times higher than on earth.

79. Fairchild Industries has entered into a joint endeavor agreement with NASA to develop an automatic orbital platform which would be deployed into low earth orbit from the Space Shuttle for periods up to six months before retrieval or in orbit servicing. Fairchild plans to lease space aboard the LEASECRAFT to manufacturers for experiments and production on a commercial basis.

80. At the same time, British Telecom's monopoly in intercontinental links was broken up, as early as 1981, when the Mercury network was set up (controlled by companies such as BP, Barclays, Cable and Wireless).

81. Each company is planning to launch two satellites positioned above the Atlantic to provide a communications service between Europe and the United States (see footnote 25). Both have already submitted their licensing applications to appropriate US and European Authorities.

82. As part of the Customs Cooperation Council work on updating and harmonizing the system of trade nomenclature, the Aircraft and Space section is being modified to include a subsection on spacecraft industry including satellites and launch vehicles. Therefore trade statistics should be available after completion of this project in January 1987.

ANNEXES

EXPENDITURE ON THE DEVELOPMENT OF TELECOMMUNICATIONS SATELLITES
(excluding launching costs, current prices in millions of european currency unit)

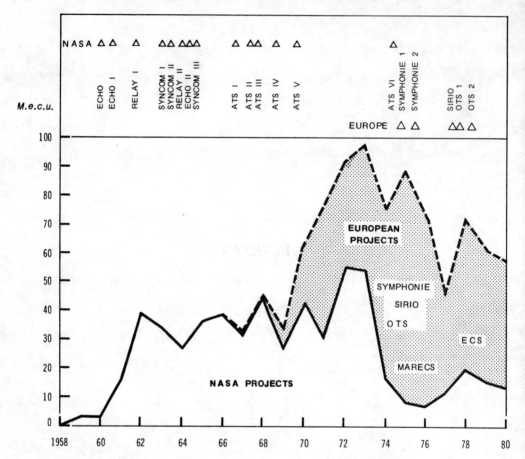

Source : *Futuribles* November, 1980.

ESA Spendings

KAU

	1982	1983	1984
General budget	89 907	92 177	93 113
Associated to general budget	35 633	52 739	84 114
Total	125 540	144 916	177 227
Scientific programmes	107 017	121 744	127 471
Earth observation programme	42 676	59 168	162 914
Telecommunications programme	228 477	261 558	289 034
STS/SPACELAB programme	75 244	97 555	126 352
STS/ARIANE programme	177 885	212 479	226 054
Total	756 839	897 420	1 109 052
Financed by third parties	83 822	79 234	59 824
Total general	840 661	976 654	1 168 876

Source: Annual ESA financial accounts.

EVOLUTION OF INTELSAT SATELLITES

INTELSAT designation	I	II	III	IV	IV-A	V	V-A	VI
Year of first launch	1965	1966	1968	1971	1975	1980	1983	1986
Prime contractor	HUGHES	HUGHES	TRW	HUGHES	HUGHES	FORD AEROSPACE	FORD AEROSPACE	HUGHES
Dimensions (Undeployed)								
Width, m	0.7	1.4	1.4	2.4	2.4	2.0	2.0	3.6
Height, m	0.6	0.7	1.0	5.3	6.8	6.4	6.4	5.3
Launch vehicles		THOR DELTA		ATLAS CENTAUR		ATLAS CENTAUR or ARIANE 1-2		STS or ARIANE 4
Spacecraf transfer orbit mass, kg	68	162	293	1 385	1 489	1 946	2 140	12 100/ 3 720
Communications paylod mass, kg	13	36	56	185	190	235	260	600
Power EOL equinox, Watts	40	75	134	460	800	1 270	1 270	2 200
Design lifetime, years	1,5	3	5	7	7	7	7	10
Rated voice channel capacity in ADR	480	480	2 400	8 000	12 000	25 000	30 000	80 000
Bandwidth, MHz	50	130	300	500	800	2 300	2 180	3 680
Antenna beam coverages								
C–Band	Toroidal northern only	Toroidal almost full earth	Despun earth cover	Despun earth cov and 2 spots steerable	Despun earth cov. and 2 hemi	3 - Axis earth cov. 2 hemi. 2 zone	3 - Axis dual pol earth cov. 2 hem. 2 zone 2 spots	Despun dual pol earth cov. 2 hem. 4 zones
Ku–Band	n.a.	n.a.	n.a.	n.a.	n.a.	2 spots steerable	2 spots steerable	2 spots steerable
L–Band	n.a.	n.a.	n.a.	n.a.	n.a.	Earth cov.	n.a.	n.a.

Source : INTELSAT.

PLANNED GEOSYNCHRONOUS COMMUNICATIONS EXISTING SATELLITES (CIVILIAN)

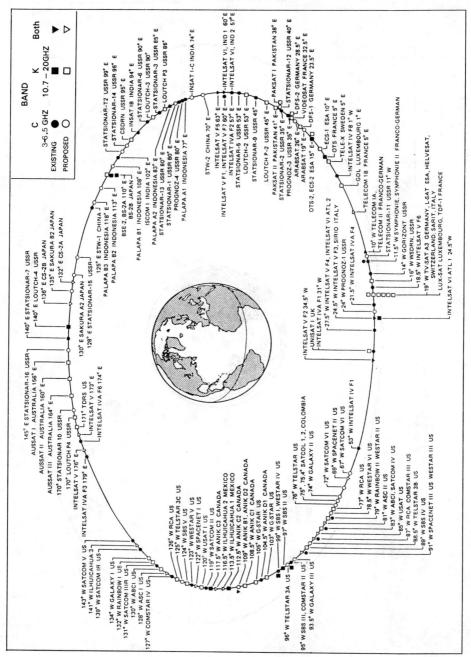

Source : Ford Aerospace and Communication Corporation (1983).

EXISTING / PLANNED GEOSYNCHRONOUS SATELLITES FOR MARITIME, METEOROLOGICAL AND MILITARY SERVICES

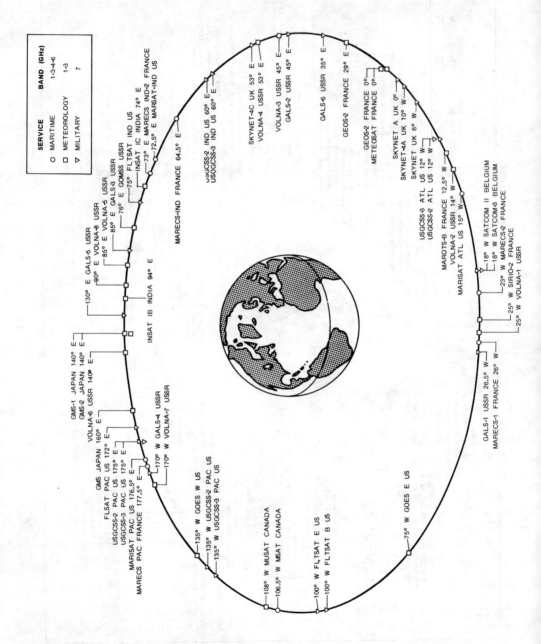

Source: Ford Aerospace and Communication Corporation, 1983.

70

GROWTH OF TRAFFIC IN HALF CIRCUITS

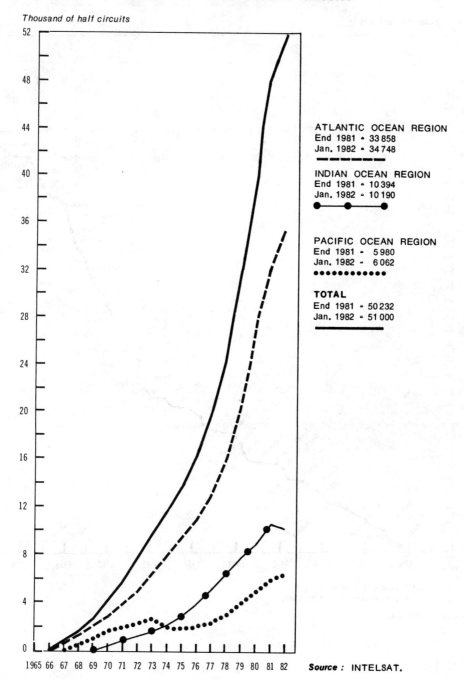

Thousand of half circuits

ATLANTIC OCEAN REGION
End 1981 • 33 858
Jan. 1982 • 34 748

INDIAN OCEAN REGION
End 1981 • 10 394
Jan. 1982 • 10 190

PACIFIC OCEAN REGION
End 1981 • 5 980
Jan. 1982 • 6 062

TOTAL
End 1981 • 50 232
Jan. 1982 • 51 000

Source : INTELSAT.

GROWTH OF INTELSAT DOMESTIC TRANSPONDER LEASES

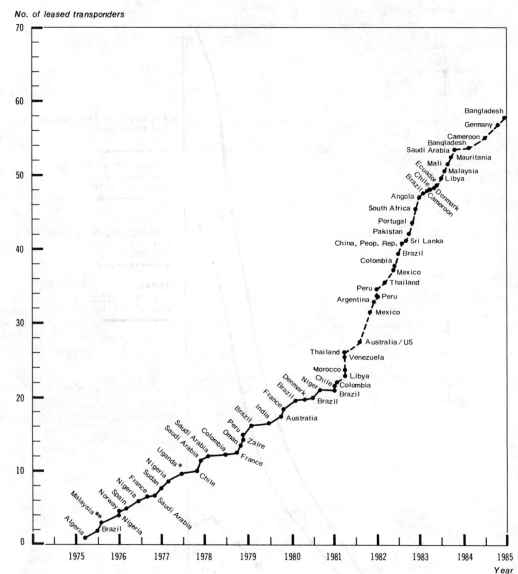

No. of leased transponders

Year

*   Terminated lease effective April 1979.
** Terminated lease effective 31 July 1980.

Source : INTELSAT.

Annex I.8

Decreasing cost of satellite systems

Year	Programme	Number of equiv. (36 MHz) transponders	Cost/transponder/year ($ million)
		Band	1982
1972	WESTAR	12 C-Band	0.65
1974	SATCOM	24 C-Band	0.50
1978	SBS	12 Ku-Band	0.93
1981	GALAXY	24 C-Band	0.37
1981	SATCOM	24 C-Band	0.26
1981	GSTAR	24 Ku-Band	0.34
1982	SPACENET	24 C-Band 12 Ku-Band	0.28
1988	FORD	24 C-Band 24 Ku-Band	0.30

Source :   Ford Aerospace and Communication Corporation.

73

Annex I 9

FAMILY OF CURRENT US LAUNCH VEHICLES

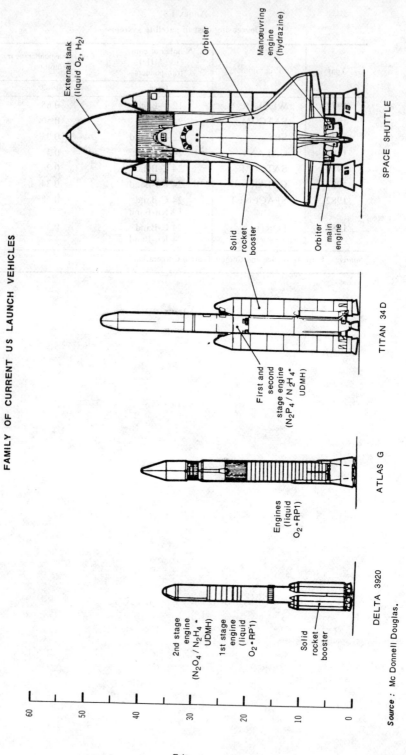

Source : Mc Donnell Douglas.

74

EUROPEAN LAUNCHERS THE ARIANE FAMILY

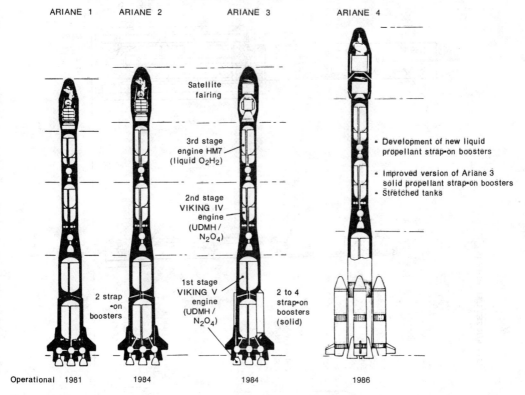

ARIANE 1   ARIANE 2   ARIANE 3   ARIANE 4

Satellite
fairing

3rd stage
engine HM7
(liquid O₂H₂)

2nd stage
VIKING IV
engine
(UDMH /
N₂O₄)

2 strap
•on
boosters

1st stage
VIKING V
engine
(UDMH /
N₂O₄)

2 to 4
strap•on
boosters
(solid)

- Development of new liquid
  propellant strap•on boosters
- Improved version of Ariane 3
  solid propellant strap•on boosters
- Stretched tanks

Operational   1981   1984   1984   1986

Source : Arianespace.

JAPANESE LAUNCHERS

Payload fairing

2nd stage engine (liquid O_2 H_2)

Solid auxiliary engine

1st stage engine (liquid O_2 H_2)

3rd stage solid motor

2nd stage liquid engine (LOX)

1st stage main engine liquid (LOX)

Satellite

3rd stage solid motor

2nd stage liquid engine (N_2O_4)

Strap-on booster (solid)

1st stage main engine (LOX)

3rd stage solid motor

2nd stage liquid engine (N_2O_4)

Strap-on booster (solid)

1st stage main engine (LOX)

Satellite fairing third stage

2nd stage

1st stage

	N I	N II	H 1	H 2
First flight	1975	1981	1988	1990
Maximum payload in geostationary orbit (tons)	0.13	0.35	0.55	2

m

50

40

30

20

10

0

Source : NASDA.

Annex I.12

Member states' contributions to the main ESA programmes in 1982

In percentage

	General budget and science	Meteosat experimental phase 2	Sirio 2	OTS	ECS	ECS phase 3 bis	Marecs A	Marecs B	ASTP[a]	L Sat phase C/D (Olympus)	Spacelab FOD[b]	Micro-gravity	ELA 2	Ariane 4
Member states														
Germany	25.57	26.25	9.00	25.00	30.68	30.42	19.08	13.29	–	–	42.00	27.57	21.00	20.79
Belgium	4.49	4.60	3.30	5.17	3.27	3.19	0.95	0.14	2.23	3.70	3.50	4.49	11.00	2.80
Denmark	2.51	2.99	–	2.90	0.33	0.74	–	–	1.84	1.30	0.50	2.51	2.75	0.15
Spain	5.04	–	0.50	–	0.17	0.53	0.95	0.34	8.30	2.60	2.80	1.00	2.50	2.50
France	21.40	25.57	14.99	20.67	25.93	26.52	11.92	5.74	–	–	15.00	15.50	59.55	52.90
Ireland	0.54	–	–	–	–	–	–	–	–	–	–	–	–	0.04
Italy	12.46	12.74	64.90	14.38	14.78	13.85	2.20	1.28	55.61	32.80	14.00	7.50	2.00	7.75
Netherlands	6.00	–	–	2.50	0.94	1.77	4.63	1.49	2.52	11.80	–	4.00	–	2.00
United Kingdom	13.75	14.37	1.83	15.86	20.15	18.46	55.81	69.89	20.51	34.30	2.00	1.35	–	3.55
Sweden	4.25	–	1.50	4.91	1.62	3.97	2.96	6.61	5.14	–	–	4.25	–	1.39
Switzerland	3.99	4.19	3.50	4.59	2.13	0.55	–	–	1.80	–	1.00	4.06	1.20	1.60
Other participants														
Austria	–	–	0.48	–	–	–	–	–	2.05	0.75	–	–	–	–
Canada	–	–	–	–	–	–	–	–	–	9.00	–	–	–	–
Norway	–	–	–	–	–	–	1.50	1.22	–	–	–	–	–	–
Other income	9.29	–	–	4.02	–	–	–	–	–	3.75	19.20	27.77	–	4.53

a A.S.T.P.: Advanced System Technology Programme.
b F.O.D.: Follow on Development (Spacelab).

Source: ESA.

77

GROWTH OF NASDA'S BUDGET

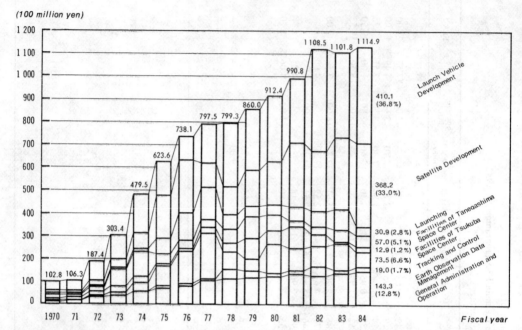

Note : The figures in parentheses are the percentages of the total budget.

● Dollar figures calculated, for information only, at rate of 240 yen to the dollar (US).

Source : NASDA.

78

WORLD MARKET SHARES FOR CIVIL TELECOMMUNICATIONS SATELLITES BY PRIME CONTRACTOR AND SUBCONTRACTOR

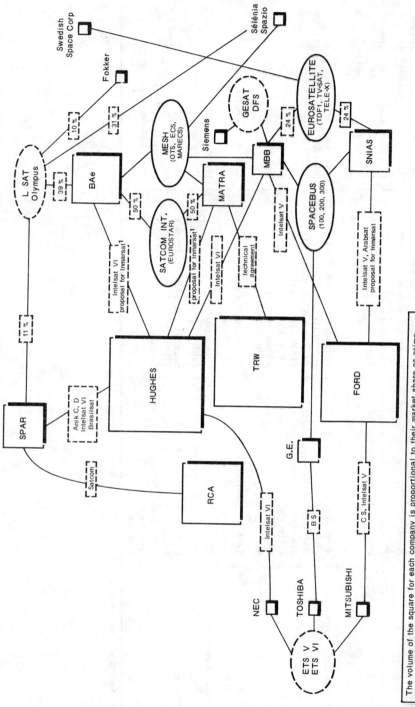

The volume of the square for each company is proportional to their market share as prime contractor for telecommunication civil satellites launched between 1980 and 1987.

1. Recently, the consortium led by BAE won the contract for the construction of the new generation of INMARSAT satellites. Hughes (payload prime contractor) and SATCOM INTERNATIONAL are members of this consortium.

Source : Euroconsult / Espace 84-85.

Annex I.15

The european space consortia: composition and main achievements

	MESH	STAR	COSMOS
Denomination	MESH: Matra, Erno, Saab, Hawker Siddeley	STAR: satellites for telecommunications applications and research	COSMOS: consortium for study and manufacturing of satellites
Constitution	October 1966: four-firms consortium December 1969: joining of FIAT	1969: creation of EST September 1970: evolution to STAR	1969 creation of "BAC" consortium 1979: creation of CESAR November 1970: evolution to COSMOS
Represented countries	Matra (F) ERNO (D) HSD (now BADG Stevenage) (UK) FIAT (now Aeritalia) (I) Fokker (NL) INTA (E) Saab-Scania (S)	Thomson-CSF and SEP (F) Dornier (D) BAC (now BADG Bristol) (UK) FIAR and Laben (I) Sener (E) LM Ericsson (S) Contraves (CH)	SNIAS and SAT (E) MBB and Siemens* (D) MSDS (RU) Selenia (I) ETCA (B) Casa (E)
Tech. ass./special commercial relations	TRW Systems (USA)	Hughes Space and Comm. (USA)	Aeronautic Ford (USA)
Main achievements	TD-1A Spacelab (enlarged consortium)	Geos 1 and 2 ISEE-B	Meteosat 1 and 2 preparation of operations
Meteosat	OTS, Marots/Marecs, ECS Telecom 1 (French programme) L-Sat (BADG Stevenage + ad hoc consortium) Hipparcos	Space Telescope (BADG Bristol) ISPM Giotto	Intelsat V (with Ford) Exosat (H-Sat)-TV-Sat/TDF (Franco-German programme)

* Siemens withdrew from space activities in 1979 except for ground stations.

Annex 1.16

Role of the consortia in ESRO/ESA projects

Period	Project	Main contractor	Consortium	Notes
1964-1968	ESRO II	HSD	(MESH)	Consortia not yet formed
	ESRO I	LCT	–	
	ESRO IV	HSD	(MESH)	
	HEOS A1	Junkers	(COSMOS)	
	HEOS A2	Junkers	(COSMOS)	
1966-74	TD	Matra	MESH	Contracts with a prime contractor representing consortium
	OTS	HSD	MESH	
	Marots platform	HSD	MESH	
	Geos	BAC	STAR	
	ISEE	Dornier	STAR	
	Cos-B	MBB	CESAR/COSMOS	
	Meteosat	SNIAS	COSMOS	
	Marots payload	MSDS	COSMOS	
After 1974	Spacelab	ERNO	–	Contracts with a prime contractor not representing (or only partially representing) a consortium
	S/L IPS	Dornier	–	
	Ariane	SNIAS	(COSMOS)	
	Exosat	MBB	Solar array BADG	
	Space Telescope	Lockheed + Perkin Elmer	PDA: Dornier	
		(EU)	Camera mod: BADG	
			(STAR)	
	ISPM	Dornier	MESH	
	ECS	BADG Stevenage	(MESH)	
	Marecs	BADG Stevenage + MSDS	–	
	Sirio 2	CNA	(STAR)	
	L-Sat	BADG Stevenage	(MESH)	
	Giotto	BADG Stevenage		
	Hipparcos	Matra		

81

EVOLUTION OF EUROPEAN INDUSTRIAL CONSORTIA

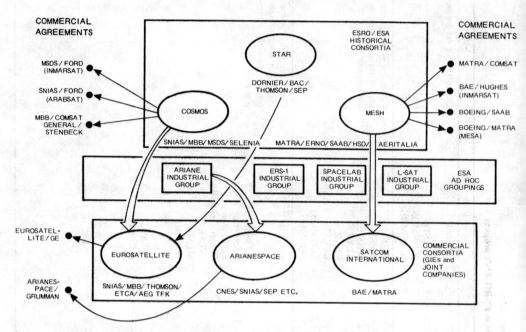

COMMERCIAL
AGREEMENTS

COMMERCIAL
AGREEMENTS

ESRO / ESA
HISTORICAL
CONSORTIA

STAR

DORNIER / BAC /
THOMSON / SEP

MSDS / FORD
(INMARSAT)

SNIAS / FORD
(ARABSAT)

MBB / COMSAT
GENERAL /
STENBECK

COSMOS

MESH

MATRA / COMSAT

BAE / HUGHES
(INMARSAT)

BOEING / SAAB

BOEING / MATRA
(MESA)

SNIAS / MBB / MSDS / SELENIA   MATRA / ERNO / SAAB / HSD /   AERITALIA

ARIANE
INDUSTRIAL
GROUP

ERS-1
INDUSTRIAL
GROUP

SPACELAB
INDUSTRIAL
GROUP

L-SAT
INDUSTRIAL
GROUP

ESA
AD HOC
GROUPINGS

EUROSATEL-
LITE / GE

EUROSATELLITE

ARIANESPACE

SATCOM
INTERNATIONAL

COMMERCIAL
CONSORTIA
(GIEs and
JOINT
COMPANIES)

ARIANES-
PACE /
GRUMMAN

SNIAS / MBB / THOMSON /
ETCA / AEG TFK

CNES / SNIAS / SEP ETC.

BAE / MATRA

Source : ESA Bulletin No. 38, May 1984.

Annex I 18

COMPANIES BUILDING NATIONAL AND REGIONAL COMMUNICATIONS SATELLITE SYSTEMS

Country System	Space-craft	RCA	Hughes	Spar	Ford	Melco	BAe Matra	SNIAS	MBB / ANT	Siemens MBB/ ANT
Canada										
Anik C	3		■							
Anik D	2			■						
United States										
WU Westar	4		■							
RCA C Band	6	■								
SBS	6		■							
AT & T Telstar	3		■							
HCI Galaxy	3		■							
GTE Gstar	4	■								
GTE Spacenet	4	■								
ASC	2	■								
RCA K Band	3	■								
Indonesia										
Palapa	2		■							
Brazil										
SBTS	2			■						
Australia										
Aussat	3		■							
Mexico										
Morelos	2		■							
India										
Insat	3				■					
Japan										
Sakura 2	2					■				
CS-3	2					■				
Arab League										
Arabsat	3							■		
Europe										
Eutelsat	5						■			
Germany										
TV Sat	1								■	
Kopernicus	2									■
France										
Telecom	4						■			
United Kingdom										
Unisat	2						■			
Total	73	19	26	4	3	4	11	3	1	2

Source : A.D. Wheelon, World Telecommunication Forum, Geneva, October 1983 and Secretariat.

83

Annex I.19

Recent performance and forecast: guided missiles and space vehicles (SIC 3761)

In millions of dollars except as noted

	1982	1983[1]	1984[2]	1985[3]	Percent Change		
					1982-83	1983-84	1984-85
Industry data							
Value of shipments	10 286	9 960	11 700	–	–3.2	17.5	–
Value of shipments (1972 $)	4 226	3 917	4 412	5 062	–7.3	12.6	14.7
Total employment (000)	99.6	107.2	108.2	110.0	7.6	0.9	1.7
Production workers (000)	35.9	41.8	44.2	48.0	16.4	5.7	8.6
Average hourly earnings ($)	12.99	13.86	14.07	–	6.7	1.5	–
Product data							
Value of shipments	8 653	8 300	9 800	–	–4.1	18.1	–
Value of shipments (1972 $)	3 526	3 242	3 665	4 303	–8.1	13.0	17.4
(1972 = 100)	244.4	255.4	262.7	–	4.5	2.9	–
Trade data							
Value of imports	0.0	39.5	0.0	0.0	–	–	–
Value of exports	716.4	442.8	375.0	400.0	–38.2	–15.3	6.7

1. Estimated except for shipments price index, exports and imports.
2. Estimated.
3. Forecast.
Source: US Department of Commerce: Bureau of the Census, Bureau of Economic Analysis, and International Trade Administration (ITA). Estimates and forecasts by ITA.

Annex I.20

Recent performance and forecast: space propulsion units and parts (SIC 3764)

In millions of dollars except as noted

	1982	1983[1]	1984[2]	1985[3]	Percent Change		
					1982-83	1983-84	1984-85
Industry data							
Value of shipments	2 222	2 050	2 350	–	–7.7	14.6	–
Value of shipments (1972 $)	825	726	796	939	–11.9	9.6	18.0
Total employment (000)	25.2	27.1	28.2	29.0	7.5	4.1	2.8
Production workers (000)	10.7	12.3	13.1	14.5	15.0	6.5	10.7
Average hourly earnings ($)	11.25	12.00	12.20	–	6.7	1.7	–
Product data							
Value of shipments	2 199	2 000	2 300	–	–9.1	15.0	–
Value of shipments (1972 $)	817	709	779	907	–13.2	9.9	16.4
(1972 = 100)	283.3	295.4	312.3	–	4.3	5.7	–
Trade data							
Value of imports	0.0	0.0	0.0	0.0	–	–	–
Valeur des exportations	1.9	51.8	45.0	50.0	–62.4	–13.1	11.1

1. Estimated except for shipments price index, exports and imports.
2. Estimated.
3. Forecast.
Source: US Department of Commerce: Bureau of the Census, Bureau of Economic Analysis, and International Trade Administration (ITA). Estimates and forecasts by ITA.

Annex I.21

Recent performance and forecast: space vehicles equipment, n.e.c. (SIC 3769)

In millions of dollars except as noted

	1982	1983[1]	1984[2]	1985[3]	Percent Change		
					1982-83	1983-84	1984-85
Industry data							
Value of shipments	1 871	2 494	2 900	–	–33.3	16.3	–
Value of shipments (1972 $)	731	925	1 026	1 146	–26.7	10.8	11.7
Total employment (000)	20.5	21.9	22.8	23.6	6.8	4.1	3.5
Production workers (000)	12.3	14.1	15.0	16.5	14.6	6.4	10.0
Average hourly earnings ($)	11.73	12.50	12.70	–	6.6	1.6	–
Product data							
Value of shipments	2 756	3 616	4 200	–	–31.2	16.2	–
Value of shipments (1972 $)	1 076	1 342	1 486	1 686	–24.7	10.7	13.5
(1972 = 100)	266.7	279.4	293.1	–	4.8	4.9	–
Trade data							
Value of imports	0.0	0.0	0.0	0.0	–	–	–
Value of exports	377.9	498.8	645.0	840.0	–32.0	–29.3	30.2

1. Estimated except for shipments price index, exports and imports.
2. Estimated.
3. Forecast.
Source: US Department of Commerce: Bureau of the Census, Bureau of Economic Analysis, and International Trade Administration (ITA). Estimates and forecasts by ITA.

ANNUAL FACTORY SHIPMENTS OF US SPACE SATELLITE COMMUNICATION SYSTEMS, 1974-1983

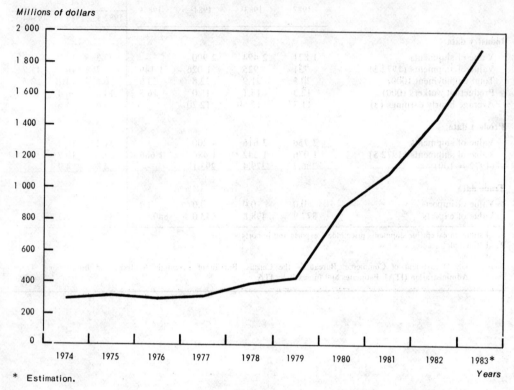

Millions of dollars

* Estimation.

Source : Electronic Industries Association, Electronic Market Data Book, Washington, 1984 and US Department of Commerce.

GROWTH OF GLOBAL SPACE ACTIVITIES TURNOVER
(for a sample of representative companies in Europe)

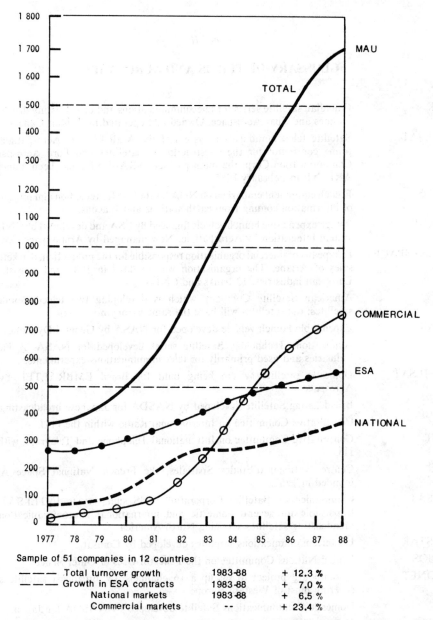

Sample of 51 companies in 12 countries

— — — Total turnover growth	1983-88	+ 12.3 %
— — — Growth in ESA contracts	1983-88	+ 7.0 %
National markets	1983-88	+ 6.5 %
Commercial markets	--	+ 23.4 %

Source : The Workload of European Space Industry. Dondi and Toussaint.
ESA Bulletin May, 1984.

Annex II

GLOSSARY OF TERMS AND ACRONYMS

ANIK Canadian satellite series for telecommunications and broadcasting developed by Hughes and Spar Aerospace. Owned and operated by Telesat Canada.

ARABSAT Satellite telecommunication system of the Arab League. Aerospatiale is the prime contractor for the construction of satellites and Ford Aerospace and Communication Corp. the main partner. ARABSAT 1 has been launched by ARIANE in February 1985.

ARGOS French equipment embarked on NOAA satellites for reception and preprocessing of information coming from earth stations and beacons.

ARIANE 3 stage expendable launch vehicle financed by ESA and developed by CNES with system integration by Aerospatiale. Now managed by Arianespace.

ARIANESPACE European commercial organisation responsible for the production, marketing and sales of Ariane. The organisation was founded in 1980 and consists of 36 European industries, 13 banks and CNES.

ASC American Satellite Company which is developing two telecommunications satellites; the satellites will have the same acronym.

ATLAS Expendable launch vehicle developed for NASA by General Dynamics.

ATS Applications Technology Satellite series developed for NASA by Fairchild Industries and used primarily for telecommunications experiments.

BRASILSAT Domestic satellite system being built for Brasil EMBRAETEL by Spar Aerospace.

BS Broadcasting Satellite developed by NASDA for Japanese broadcasting.

CCIR Consultative Committee on International Radio within the ITU.

CCITT Consultative Committee on International Telephone and Telegraph within the ITU.

CNES Centre National d'Etudes Spatiales, the French National Space Agency, founded in 1962.

COMSAT Communication Satellite Corporation - US signatory to INTELSAT. It is responsible for several domestic and international telecommunications and broadcasting satellites servicing North America.

COMSTAR US telecommunications satellites developed by Comsat.

COPUOS United Nations Committee on the Peaceful Uses of Outer Space.

CORONET Luxembourg project to set up a 16 channels TV broadcast satellite system covering most of Western Europe.

CS Domestic Communications Satellite developed by NASDA for Japan.

DARPA Defense Advanced Research Projects Agency, US Department of Defense. Performs advanced research in several areas, including space technology.

88

DBS	Direct Broadcast Satellites used primarily for direct TV broadcasting to homes.
DELTA	3 stage expendable launch vehicle developed for NASA by McDonnell Douglas.
DFS/ KOPERNIKUS	Satellite developed for the German Federal Post Office by a consortium led by Siemens (including AEG, Standard Elektrik Lorenz et MBB).
DSCS	Defense Satellite Communication System. US military communications satellites developed by TRW and General Electric.
EARLYBIRD	US telecommunications satellite launched in 1965, renamed INTELSAT I.
EARTHNET	European network for the acquisition, preprocessing and dissemination of remote sensing satellite data.
ECS	European Communications Satellites developed for regional telecommunications within Europe under direction from ESA. System integration done by British Aerospace and operations by EUTELSAT. The first satellite was launched in 1983.
ELV	Expendable Launch Vehicle such as ARIANE or DELTA which can be used once and then disintegrates in space.
ERS 1	Japanese Earth Resource Satellite. Active sensing satellite designed for the development of synthetic aperture radar technology. Scheduled for launch in the early 1990s. Also the first ESA remote sensing satellite to be launched in 1987.
ESA	European Space Agency founded in 1975 and consisting of 11 member nations and 2 associate members. Canada is co-operating with the organisation under an agreement of co-operation.
ESRO/ELDO	European Space Research Organisation and European Launch Development Organisation which were the predecessors to ESA.
ETS	Engineering Test Satellites developed by NASDA. The first of the ETS series was launched in 1975 and the fifth, ETS V, will be launched in 1987.
EUMETSAT	European Meteorological Satellite Organisation. The agreement founding the Organisation was opened for signature on 24th May 1983. Its aim is to set up and manage a European operational system of weather satellites.
EURECA	European Retrievable Carrier. Free flyer platform derived from the Spacelab pallet. First launch on the Shuttle scheduled for 1987.
EUTELSAT	European Telecommunications Satellite Organisation. The Agreement creating the organisation was opened for signature on 15th July 1982. This organisation will replace the interim EUTELSAT created in 1977. At present 20 countries in Europe have signed the agreement. EUTELSAT has three main objectives: creation and operation of a European regional satellite telecommunications system, provision of space segment for Eurovision and constitution of a satellite multiservice system.
FLEETSATCOM	US military satellites developed by TRW.
FORDSATCOM	US telecommunications and broadcasting satellite developed by Ford Aerospace. First launch in 1987.
GALAXY	US telecommunications satellites developed by Hughes. First launch in 1983.
GIOTTO	Astronomical probe developed by ESA to study the Halley Comet. Launched in 1985.
GMS	Japanese meteorological satellites. First launch in 1977.

GOES	Geostationnary Operational Environmental Satellites. US meteorological satellites.
GSO	Geostationary Orbit which is an equatorial orbit suitable for fixed telecommunications satellites. A finite number of slots are available for spacecraft. The ITU is the forum where countries co-operate to regulate the use of the GSO for satellite communication.
G-STAR	US telecommunications and broadcasting satellites developed by GTE and RCA. First launch in 1986.
H-I	Expendable launch vehicle being developed by NASDA.
H-II	Expendable launch vehicle planned to be developed by NASDA to place heavier satellites in orbit than the H-I.
HIPPARCOS	Scientific satellite developed by ESA for stars observation which will be launched in 1988.
INMARSAT	International Organisation created in 1979. It is charged with the definition, procurement and management of a world satellite maritime telecommunications system (39 countries affiliated).
INTELSAT	International Telecommunications Satellite Organisation operating 17 satellites positioned over the Atlantic, Pacific and Indian Oceans which are used primarily for transnational communications, but can also be leased for domestic purposes.
ISEE	Observation satellite designed to study the magnetosphere and the sun. Developed in cooperation by NASA and ESA.
ISPM	International Solar Polar Mission. Scientific experiments using in particular the GIOTTO probe.
ITALSAT	Italian national telecommunications and broadcasting satellite to be launched in 1987.
ITU	United Nations International Telecommunication Union.
IUS	Inertial Upper Stage for satellites launched by the Shuttle.
JEA	Joint Endeavor Agreement between NASA and US corporations for materials processing experiments aboard the Shuttle.
LANDSAT	Remote sensing satellites and services operated by NOAA. LANDSAT D and D' spacecrafts carry a multispectral scanner and a high definition thematic mapper.
LEASAT	Commercial communications satellite system developed by Hughes and fully leased by the US military.
LEO	Low Earth Orbit used by observation satellites.
MARECS	Maritime telecommunications satellites developed by ESA.
MARISAT	Maritime communications satellites developed for Comsat by Hughes.
METEOSAT I and II	Meteorological satellites developed by ESA and launched in 1977 and 1981 to provide half hourly picture of earth and clouds cover. METEOSAT operational program began in 1983.
MORELOS	Telecommunications satellite project for Mexico.
MOS 1	Marine Observation Satellite. First Japanese domestic earth observation satellite. To be launched by an N-II vehicle in the Summer of 1986 and operated by NASDA.
MSAT	Mobile Satellite. A proposed new communication system that will bring two ways mobile service by using a satellite as a relay station in space. Back up for Canada

	MSAT System (led by Telesat) is expected to be provided by a similar satellite serving the US. The Canadian system in turn will provide back up for the US system.
N-I & II	Expendable launch vehicles developed for NASDA by Mitsubishi Heavy Industries (the main contractor).
NASA	National Aeronautics and Space Administration of the US founded in 1958.
NASDA	National Space Development Agency of Japan founded in 1969.
NAVSTAR GPS	(Global Positionning System). US communications and navigation satellites constructed by Rockwell for the armed forces.
NOAA	National Oceanographic and Atmospheric Administration of the US Department of Commerce. Responsible for LANDSAT programme.
NTT	Refers to Nippon Telegraph and Telephone Public Corporation for Japanese communications.
OLYMPUS	Large telecommunications platform being developed by ESA which will be launched in 1986. Prime contractor British Aerospace.
ORION	Private company envisaging the launching of two transatlantic telecommunications satellites for services between US and Europe. Project under consideration by FCC (Federal Communication Commission).
OTS	Orbital Test Satellite developed by ESA for telecommunications experiments within Europe launched in 1978.
PALAPA	Indonesian telecommunications satellites developed by Hughes.
PAM	Payload Assist Module used to boost satellites into GSO from the main launch vehicle.
PROGRESS	USSR automatic cargo spacecraft.
PROTON	Expendable USSR launch vehicle.
PTT	Post Telephone and Telegraph organisations within European countries.
RADARSAT	Remote sensing satellite equipped with a multiple beam radar to be placed in polar orbit early in the next decade. Canada is the main contributor to the programme. Other contributors include the US and UK.
RCASat	Radio Corporation of America telecommunications and broadcasting satellite series.
RMS-Canadarm	Remote Manipulator System developed by Spar Aerospace for use on the US Shuttle.
SALYUT	USSR space station.
SATCOM	Telecommunications and broadcasting satellites developed for US domestic services by RCA.
SBS	Satellite Business Services which is a joint venture between Aetna, IBM and Comsat offering private satellite services within the US using their own satellites under the same acronym. Comsat recently sold its share.
SCOUT	Small expendable launch vehicle developed for NASA by LTV.
SIRIO	Telecommunication satellite developed for Europe by Italy.
SKYLAB	US space laboratory, placed in orbit in the early 70s and now disintegrated.
SPACELAB	Space laboratory which fits into the Shuttle's cargo bay and was developed as part of a cooperative project with ESA and NASA. The first launch aboard the Shuttle was on 28th November 1983, for 10 days' flight.
SPACENET	Telecommunications satellites for the US developed by GTE. First launch in 1984.

SPAS	Shuttle Pallet Satellite. A deployable carrier developed by MBB and first flown aboard the space shuttle in 1983.
SPOT	Remote sensing satellite project developed by CNES.
SPOT-IMAGE	Semi-private French organisation responsible for production, operations, marketing and sales of Spot services.
SPOTNET	Telecommunications satellites for US developed by National Exchange Corp. First launch in 1986.
STS	US Space Transportation System commonly referred to as the space Shuttle. A reusable system and the primary US launch vehicle.
SYMPHONIE A and B	Experimental telecommunications satellites developed by France and Germany. Launched in 1974 and 1975.
SYNCOM	First telecommunications satellite launched by NASA in 1963.
TCI	Transpace Carriers Inc. A US Company taking over the marketing and launch operations of the Delta ELV.
TDFl/TVSat	Matching broadcasting and telecommunications satellites developed jointly by SNIAS and MBB for France and Germany.
TDMA	Time Division Multiple Access (type of link between transmitters and the earth).
TDRS	Tracking Data and Relay Satellite System. Developed by NASA to replace the existing base network of tracking stations to support the tracking telemetry and data acquisition needs of US space missions in LEO.
TELECOM	Telecommunications and broadcasting satellites for France developed by Matra and CNES. The first launch took place in 1984.
TELEX	Telecommunications satellites developped by Swedish Space Corp. and built by the Eurosatellite consortium (led by Aerospatiale).
TELSTAR	Telecommunications satellites for AT&T developed by Hughes.
TIROS	Family of US polar orbiting meteorological satellites.
TITAN	Expendable launch vehicle developed for US Department of Defense by Martin Marietta.
UNISAT	Direct broadcasting and telecommunications satellites for the United Kingdom developed by British Aerospace. First launch after 1987.
U-SAT	Telecommunications satellites for the US developed by US Satellite Services.
VIKING	Swedish scientific spacecraft based on a Boeing satellite platform design MESA. Scheduled for launch in 1985.
WESTAR	Telecommunications satellites for Western Union developed by Hughes.
WMO	United Nations World Meteorological Organisation.

OECD SALES AGENTS
DÉPOSITAIRES DES PUBLICATIONS DE L'OCDE

ARGENTINA - ARGENTINE
Carlos Hirsch S.R.L., Florida 165, 4° Piso (Galería Guemes)
1333 BUENOS AIRES, Tel. 33.1787.2391 y 30.7122
AUSTRALIA - AUSTRALIE
D.A. Book (Aust.) Pty. Ltd.
11-13 Station Street (P.O. Box 163)
MITCHAM, Vic. 3132. Tel. (03) 873 4411
AUSTRIA - AUTRICHE
OECD Publications and Information Center
4 Simrockstrasse 5300 Bonn (Germany). Tel. (0228) 21.60.45
Local Agent/Agent local :
Gerold and Co., Graben 31, WIEN 1. Tel. 52.22.35
BELGIUM - BELGIQUE
Jean De Lannoy, Service Publications OCDE
avenue du Roi 202, B-1060 BRUXELLES. Tel. 02/538.51.69
CANADA
Renouf Publishing Company Limited/
Éditions Renouf Limitée Head Office/Siège social - Store/Magasin :
61, rue Sparks Street,
OTTAWA, Ontario KIP 5A6. Tel. (613)238-8985. 1-800-267-4164
Store/Magasin: 211, rue Yonge Street,
TORONTO, Ontario M5B 1M4. Tel. (416)363-3171
Regional Sales Office/
Bureau des Ventes régional :
7575 Trans-Canada Hwy., Suite 305,
SAINT-LAURENT, Québec H4T 1V6. Tél. (514)335-9274
DENMARK - DANEMARK
Munksgaard Export and Subscription Service
35, Nørre Søgade
DK 1370 KØBENHAVN K. Tel. +45.1.12.85.70
FINLAND - FINLANDE
Akateeminen Kirjakauppa
Keskuskatu 1, 00100 HELSINKI 10. Tel. 65.11.22
FRANCE
OCDE, 2, rue André-Pascal, 75775 PARIS CEDEX 16
Tel. (1) 45.24.82.00
Librairie/Bookshop : 33, rue Octave-Feuillet,
75016 PARIS. Tél. (1) 45.24.81.67 ou (1) 45.24.81.81
Principal correspondant :
13602 AIX-EN-PROVENCE : Librairie de l'Université.
Tél. 42.26.18.08
GERMANY - ALLEMAGNE
OECD Publications and Information Center
4 Simrockstrasse 5300 BONN Tel. (0228) 21.60.45
GREECE - GRÈCE
Librairie Kauffmann, 28 rue du Stade,
ATHÈNES 132. Tel. 322.21.60
HONG-KONG
Government Information Services,
Publications (Sales) Office,
Beaconsfield House, 4/F.,
Queen's Road Central
ICELAND - ISLANDE
Snaebjörn Jönsson and Co., h.f.,
Hafnarstraeti 4 and 9, P.O.B. 1131, REYKJAVIK.
Tel. 13133/14281/11936
INDIA - INDE
Oxford Book and Stationery Co. :
NEW DELHI-1, Scindia House. Tel. 45896
CALCUTTA 700016, 17 Park Street. Tel. 240832
INDONESIA - INDONÉSIE
PDIN-LIPI, P.O. Box 3065/JKT., JAKARTA, Tel. 583467
IRELAND - IRLANDE
TDC Publishers - Library Suppliers
12 North Frederick Street, DUBLIN 1 Tel. 744835-749677
ITALY - ITALIE
Libreria Commissionaria Sansoni :
Via Lamarmora 45, 50121 FIRENZE. Tel. 579751/584468
Via Bartolini 29, 20155 MILANO. Tel. 365083
Sub-depositari :
Ugo Tassi
Via A. Farnese 28, 00192 ROMA. Tel. 310590
Editrice e Libreria Herder,
Piazza Montecitorio 120, 00186 ROMA. Tel. 6794628
Agenzia Libraria Pegaso,
Via de Romita 5, 70121 BARI. Tel. 540.105/540.195
Agenzia Libraria Pegaso, Via S. Anna dei Lombardi 16, 80134 NAPOLI.
Tel. 314180.
Libreria Hoepli, Via Hoepli 5, 20121 MILANO. Tel. 865446
Libreria Scientifica, Dott. Lucio de Biasio "Aeiou"
Via Meravigli 16, 20123 MILANO Tel. 807679
Libreria Zanichelli
Piazza Galvani 1/A, 40124 Bologna Tel. 237389
Libreria Lattes, Via Garibaldi 3, 10122 TORINO. Tel. 519274
La diffusione delle edizioni OCSE è inoltre assicurata dalle migliori librerie nelle
città più importanti.
JAPAN - JAPON
OECD Publications and Information Center,
Landic Akasaka Bldg., 2-3-4 Akasaka,
Minato-ku, TOKYO 107 Tel. 586.2016
KOREA - CORÉE
Pan Korea Book Corporation,
P.O. Box n° 101 Kwangwhamun, SÉOUL. Tel. 72.7369
LEBANON - LIBAN
Documenta Scientifica/Redico,
Edison Building, Bliss Street, P.O. Box 5641, BEIRUT.
Tel. 354429 - 344425

MALAYSIA - MALAISIE
University of Malaya Co-operative Bookshop Ltd.
P.O. Box 1127, Jalan Pantai Baru
KUALA LUMPUR. Tel. 577701/577072
THE NETHERLANDS - PAYS-BAS
Staatsuitgeverij, Verzendboekhandel,
Chr. Plantijnstraat 1 Postbus 20014
2500 EA S-GRAVENHAGE. Tel. nr. 070.789911
Voor bestellingen: Tel. 070.789208
NEW ZEALAND - NOUVELLE-ZÉLANDE
Publications Section,
Government Printing Office Bookshops:
AUCKLAND: Retail Bookshop: 25 Rutland Street,
Mail Orders: 85 Beach Road, Private Bag C.P.O.
HAMILTON: Retail: Ward Street,
Mail Orders, P.O. Box 857
WELLINGTON: Retail: Mulgrave Street (Head Office),
Cubacade World Trade Centre
Mail Orders: Private Bag
CHRISTCHURCH: Retail: 159 Hereford Street,
Mail Orders: Private Bag
DUNEDIN: Retail: Princes Street
Mail Order: P.O. Box 1104
NORWAY - NORVÈGE
Tanum-Karl Johan a.s
P.O. Box 1177 Sentrum, 0107 OSLO 1. Tel. (02) 80.12.60
PAKISTAN
Mirza Book Agency, 65 Shahrah Quaid-E-Azam, LAHORE 3.
Tel. 66839
PORTUGAL
Livraria Portugal, Rua do Carmo 70-74,
1117 LISBOA CODEX. Tel. 360582/3
SINGAPORE - SINGAPOUR
Information Publications Pte Ltd,
Pei-Fu Industrial Building,
24 New Industrial Road N° 02-06
SINGAPORE 1953, Tel. 2831786, 2831798
SPAIN - ESPAGNE
Mundi-Prensa Libros, S.A.
Castelló 37, Apartado 1223, MADRID-28001, Tel. 275.46.55
Libreria Bosch, Ronda Universidad 11, BARCELONA 7.
Tel. 317.53.08, 317.53.58
SWEDEN - SUÈDE
AB CE Fritzes Kungl Hovbokhandel,
Box 16 356, S 103 27 STH, Regeringsgatan 12,
DS STOCKHOLM. Tel. 08/23.89.00
Subscription Agency/Abonnements:
Wennergren-Williams AB,
Box 30004, S104 25 STOCKHOLM. Tel. 08/54.12.00
SWITZERLAND - SUISSE
OECD Publications and Information Center
4 Simrockstrasse 5300 BONN (Germany). Tel. (0228) 21.60.45
Local Agents/Agents locaux
Librairie Payot, 6 rue Grenus, 1211 GENÈVE 11. Tel. 022.31.89.50
TAIWAN - FORMOSE
Good Faith Worldwide Int'l Co., Ltd.
9th floor, No. 118, Sec. 2,
Chung Hsiao E. Road. TAIPEI. Tel. 391.7396/391.7397
THAILAND - THAILANDE
Suksit Siam Co., Ltd., 1715 Rama IV Rd,
Samyan, BANGKOK 5. Tel. 2511630
TURKEY - TURQUIE
Kültur Yayinlari Is-Türk Ltd. Sti.
Atatürk Bulvari No : 191/Kat. 21
Kavaklidere/ANKARA. Tel. 17 02 66
Dolmabahce Cad. No : 29
BESIKTAS/ISTANBUL. Tel. 60 71 88
UNITED KINGDOM - ROYAUME-UNI
H.M. Stationery Office,
P.O.B. 276, LONDON SW8 5DT.
(postal orders only)
Telephone orders: (01) 622.3316, or
49 High Holborn, LONDON WC1V 6 HB (personal callers)
Branches at: EDINBURGH, BIRMINGHAM, BRISTOL,
MANCHESTER, BELFAST.
UNITED STATES OF AMERICA - ÉTATS-UNIS
OECD Publications and Information Center, Suite 1207,
1750 Pennsylvania Ave., N.W. WASHINGTON, D.C.20006 - 4582
Tel. (202) 724.1857
VENEZUELA
Libreria del Este, Avda. F. Miranda 52, Edificio Galipan,
CARACAS 106. Tel. 32.23.01/33.26.04/31.58.38
YUGOSLAVIA - YOUGOSLAVIE
Jugoslovenska Knjiga, Knez Mihajlova 2, P.O.B. 36, BEOGRAD.
Tel. 621.992

Les commandes provenant de pays où l'OCDE n'a pas encore désigné de dépositaire peuvent être adressées à :
OCDE, Bureau des Publications, 2, rue André-Pascal, 75775 PARIS CEDEX 16.

Orders and inquiries from countries where sales agents have not yet been appointed may be sent to:
OECD, Publications Office, 2, rue André-Pascal, 75775 PARIS CEDEX 16.

69072-10-1985

OECD PUBLICATIONS, 2, rue André-Pascal, 75775 PARIS CEDEX 16 - No. 43375 1985
PRINTED IN FRANCE
(70 85 03 1) ISBN 92-64-12772-0